INVENTING REALITY

The Wiley Science Editions

INVENTING REALITY

PHYSICS AS LANGUAGE

Bruce Gregory

WILEY SCIENCE EDITIONS
JOHN WILEY & SONS, INC.
New York · Chichester · Brisbane · Toronto · Singapore

For Werner

Library of Congress Cataloging-in-Publication Data

Gregory, Bruce.
  Inventing Reality: Physics as Language.

(Wiley science editions)
Bibliography: p.
  1. Physics.   2. Physics—History.   3. Quantum
theory.   4. Space and time.   I. Title.   II. Series
QC28.G74 1988   530   88-14242
ISBN 0-471-61388-6
ISBN 0-471-52482-4 (pbk)

Printed in the United States of America

90 91 10 9 8 7 6 5 4 3 2 1

PREFACE

Physics has been so immensely successful that it is difficult to avoid the conviction that what physicists have done over the past 300 years is to slowly draw back the veil that stands between us and the world as it really is—that physics, and every science, is the discovery of a ready-made world. As powerful as this metaphor is, it is useful to keep in mind that it *is* a metaphor, and that there are other ways of looking at physics and at science in general. Ways that may prove even more illuminating than the "obvious" view.

This book tells the story of how physicists invented a language in order to talk about the world. In this sense, the book is not "about" physics. Although I hope the reader will discover something about physics, my purpose is not to try to explain the discipline, but to explore the relationship between language and the world. Physicists use a very precise language, and this precision gives us an opportunity to see more clearly than is otherwise possible just how much of what we find in the world is the result of the way we talk about the world.

The physics discussed in this book is sometimes called *fundamental physics,* but this understandably offends physicists who feel what they are doing is every bit as fundamental. Much of fundamental physics is also called *high-energy physics* or *particle physics*. Other branches of physics have had a much greater effect on our lives. For example, solid-state physics has had an immense impact as a result of the development of transistors and integrated circuits. The physics whose history I trace is concerned with understanding the ultimate constituents of matter and the nature of the forces through which these constituents interact. It attempts to answer the question, What is the world made of, and how does it work? These are clearly fundamental questions, if not the only fundamental questions. I chose to focus on this branch because I think it has something particularly valuable to tell us about the general quest to understand the "furniture of the universe"—the stuff we think of as real—and the role that language plays in creating and supporting this conviction.

I use the word *language* throughout the book, and it is helpful to know what I mean by it. I call any symbolic system for dealing with the world a language. The language of physics is mathematics, and I refer to particular domains of mathematics as the language of classical physics or the language of quantum mechanics. I also refer to the vocabulary of Newtonian physics, where I mean the Newtonian approach to dealing with physical situations in terms of forces and changing position with time. I take this approach to make it clear that physics *is* a language, a way of talking about the world. As to what else we can say about physics, this is something the book sets out to explore.

The story I tell resembles physics itself in that it ruthlessly pares away the sort of detail historians rightly consider essential in understanding a place and time. The obligatory details and caveats of a truly historical exploration would obscure the story I want to tell. I have tried to be historically accurate, but I have only touched on the high points, and the reader who wishes to savor the richness of the history is referred to works recommended in the bibliography at the end of the book. I have also pared away a great deal of physics in the effort to make what is central to my discussion more intelligible; more detailed discussions of the science are recommended in the bibliography.

The reader is also entitled to know where the demarcation lies between what is broadly accepted by the community of physicists and the views that may be mine alone. By and large, the story as it is developed here is not controversial. For example, the interpretation of quantum mechanics given here follows the lines laid down by the orthodox or "Copenhagen" interpretation accepted by the great majority of physicists. There are other ways of understanding quantum mechanics, but in each case they are held by an extremely small, and in some cases dwindling, minority. They are fun to think about, but

I believe the reader without a strong background in physics more likely would be confused than enlightened by a discussion of these esoteric interpretations. Again the bibliography points to works that develop these ideas in greater detail.

I owe my recognition of the way language commits us to what is real to the writings of W.V.O. Quine. My appreciation of the implications of pragmatism was greatly enriched and deepened by the writings of Richard Rorty.

It is a pleasure to acknowledge the people who made this book possible. I benefited enormously from many conversations throughout the years with Bob and Holly Doyle that shaped and sharpened my understanding of the issues raised in the book. Barbara Clark, Steve and Susan Gross, Linda Houck, James Jones and Gail Hughes, Ken and Linda Schatz, and Susan Thomas provided continuing support and encouragement. In addition, the Schatzes introduced me to my agent, Michael Snell, who besides placing the book gave me trenchant advice on how to write clearly and effectively.

Myron Lecar read an early version of the manuscript, and Holly Doyle, Don Lautman, Lynn Margulis, Matthew Schneps, Edward Tripp, and Angela von der Lippe read later versions; they provided helpful suggestions and badly needed encouragement. Owen Gingerich, Lawrence Krauss, and Alan Lightman read parts of later versions, pointing out ambiguities and mistakes and providing helpful guidance. I also benefited greatly from the remarks of several anonymous readers. Suggestions and queries from my editor David Sobel resulted in a much more intelligible book than I would have otherwise written. The book would have been much less accurate and far more obscure were it not for several careful readings of the manuscript by Irwin Shapiro. Any incoherencies or muddles that remain do so despite his admonitions. I deeply appreciate the time he gave to the project despite his unreasonably busy schedule. I am grateful to Marian Shapiro for providing the impetus to seek a publisher for the book.

My wife, the poet Gray Jacobik, provided constant support, encouragement, and unfailingly perspicacious editorial advice. Her keen understanding of the ways in which language works was indispensable to developing the approach I take in this book. I owe more to her than I can possibly express. I owe my appreciation of the immense power of the myth of "is" to Werner Erhard's relentless commitment to making a difference in my life. Absent his unremitting efforts to uncover the role of speaking in shaping experience, this book never would have been written.

CONTENTS

To him who is a discoverer in this field, the products of his imagination appear so necessary and natural that he regards them, and would like to have them regarded by others, not as creations of thought but as given realities.

ALBERT EINSTEIN

PROLOGUE

It is not nature that is economical but science.[1]

MAX BORN

Stand at the foot of a tall building, point a camera upward, and take a picture. The picture will look badly distorted—it could hardly be called accurate. Yet there is nothing wrong with the camera, and cameras normally do not lie. We have a convention about how photographs of buildings "ought" to look, and this photograph violates that convention. The apparent distortion in the photograph tells something about what we "really" see. The visual world is not a faithful reflection of the images on the retinas of our eyes but a world somehow constructed out of such images.

The American psychologist Adelbert Ames constructed a room that appears to be normal when viewed from one perspective but which, in fact, is far from normal. For example, when people walk from one side of a distant wall to another, they also move farther from the viewer. Since they are moving away, the image on the retina of the viewer's eyes grows smaller. Normally, without any awareness on our part, we interpret this decrease in size in such a way that we see the people as normal-sized but far away. This room, however, is constructed so that the clues we ordinarily use to assess size and distance are misleading. By a trick of perspective, the distant wall seems to be perpendicular to the line of sight. Accepting this perspective, we see the people, not as farther away, but as smaller; they appear to shrink as they move from

1

one side of the room to the other and to grow larger as they move back again. The illusion is almost as powerful in a photograph. Even though we know that the changing size of people in the Ames room is an illusion, there is nothing we can do about it—we continue to see the same bizarre world. How much of what we see is similarly an "optical illusion"—an interpretation fabricated from our interaction with the world?

For a dog or a cat, no world at all seems to be revealed by photograph albums and television sets—our pets scarcely pay any attention to these images. The worlds portrayed by photographs, television, and movies are created by our interpretations—in this case, interpretations where language seems to have an important role to play—"The picnic with my sister and her family last Fourth of July." It is not difficult to see that the world revealed by television and newspapers is largely shaped by language. To see the world in terms of Muslim fundamentalists, Marxist guerrillas, or capitalist imperialists is certainly to see a world shaped by language.

We also know that *how* we say things is important. Women are different from girls, and homes can be different from houses. We can even create things by saying certain words. When the umpire says, "Safe!" he creates a score. When the foreman of the jury says, "Guilty as charged," she creates a felon. When two people standing before a cleric say, "I will," they create a marriage.

It may be harder to see the role that language plays in shaping our motives and emotions, but it seems true that most of what we call our feelings are interpretations we place on the bodily sensations we have learned to identify with anger or boredom or love. In this sense, the dog down the street may be ferocious, but it seems unlikely that it is angry in the way someone might be angry as a result of the way the IRS handled her tax case. To the extent that our behavior is shaped by our motives, feelings, and emotions, it seems fair to say that we live in a world structured, at least to some degree, by language.

When we consider the physical world, we have a much harder time seeing the role language plays. It seems obvious that there is a physical world quite independent of what we say or do about it. No matter how firmly someone believes he can fly simply by flapping his arms, it is unwise to for him to step off the roof of a tall building.

No matter how convinced a Buddhist is that the world is an illusion, she invariably leaves a room by walking through the doorway rather than through a wall. How does this physical world, which seems so impervious to wishes and desires, relate to the worlds shaped by language? One way to pursue this question is to look at the history of how we came to know the things we know about the physical world.

We normally think of science as the discovery of the facts about the natural world and the laws that govern its behavior, that is, we view science as the uncovering of an already-made world. In this book, we will follow another course. We will trace the history of physics as the evolution of a language—as the invention of new vocabularies and new ways of talking about the world. Concentrating on the language physicists use to talk about the world will establish a perspective vitally important for understanding the development of physics in the twentieth century. But even more important, tracing the development of physics will provide a powerful way of looking at the much broader question of how language hooks up with the world.

Although it may be surprising at first, we will find that physics is really not about making accurate pictures of the world. If you go to an art gallery and ask yourself which of the paintings are realistic and which are abstract, and why, you will discover that realism in painting is largely a convention. A physicist is no more engaged in painting a "realistic" picture of the world than a "realistic" painter is. For a physicist, a realistic picture is far too complex to be useful as a tool, and physics is about fashioning tools.

In many ways, physics resembles abstract painting more than it does photography. The world of physics is a world of hard edges and abstraction—a mathematical world as austere and as beautiful as a painting by Mondrian. A world in which the creativity and imagination of human beings is every bit as important as they are in music or painting. We will follow the story of men and women as they invent a language to empower themselves in one dimension of the endless human project of learning to deal with the world—a project that can be traced from the caves of Lascaux to the tunnels of Fermilab.

In the Beginning Was the Word . . .

It seems that the human mind has first to construct forms independently before we can find them in things. Kepler's marvelous achievement is a particularly fine example of the truth that knowledge cannot spring from experience alone, but only from the comparison of the inventions of the mind with observed fact.[2]

ALBERT EINSTEIN

Since its beginnings, science has been an assault on common sense. The Greek astronomer Aristarchus of Samos first argued in the third century B.C. that the earth moves around the sun rather than what we all *seem* to see—the sun moving around the earth. We do not know what led him to this conclusion, but it might have been his estimate of the size of the moon (in agreement with modern values) and the sun (much smaller than modern values). Even though the value Aristarchus calculated for the size of the sun was much too small by our standards, it was still larger than the size he calculated for the earth, and he may have reasoned the smaller object should move rather than the larger. In any case, Aristarchus's arguments were so unpersuasive that they were forgotten for almost a thousand years. So much for the better mousetrap theory.

Aristarchus's arguments failed to carry the day in the face of the apparently overwhelming evidence that the earth does *not* move. After all, we do not *feel* any movement. There are no immense winds such as would seem to be called for if the planet were turning and dragging the air behind it. Furthermore, when an object is thrown straight up, it returns close to the point where it was launched rather than some distance away, as it would seem to have to if the earth were moving beneath the object. In other words, the notion that the earth moves is wildly at odds with the evidence of our senses. Although we now "know" the earth "really" moves around the sun, most of us still talk as though the sun moves and not the earth ("sunrise" and "sunset," for example). The story of the modern scientific worldview begins with how we came to "know" the earth moves and not the sun.

The Polish astronomer Nicholas Copernicus, working at the beginning of the sixteenth century, was a key figure in the development of the modern description of the world. In many respects, including his lack of concern over the precise agreement between theory and observations, he was closer in spirit to the ancients, but his view of the relationship among the earth, sun, moon, and planets is substantially the one we hold today. Although elements of Copernicus's picture were earlier suggested by Aristarchus, they were virtually forgotten until Copernicus revived them.

For the intervening 1,400 years the accepted picture of the world was based on the fourth-century B.C. physics of Aristotle and the second-century A.D. astronomy of Ptolemy. One of the earliest beliefs of human beings, and one that still enjoys considerable popularity today, is the idea the stars control the fate of men and women. If we think that the stars and planets rule our destinies, then it is a good idea to know how the stars and planets behaved in the past and how they are going to behave in the future. The Greek-Egyptian astronomer Ptolemy developed a model of the behavior of the heavens in the second century that made such retrospection and prediction possible.

Ptolemy's model of the solar system places the earth in the center of the universe. The sun, moon, and planets circle the earth, and the stars are fixed to a giant sphere that lies beyond. The hard part of describing motions in the sky is reproducing the complex behavior of the planets. The stars move in smooth uniform paths across the sky, just as the sun does, but the planets move in different patterns. They move against the background of the stars as does the sun. Sometimes the planets move in the same direction as the sun, and sometimes they seem to stop moving against the background of stars and to reverse direction. After moving in the reverse direction for a while, they stop again and resume traveling in the original direction. The brightness of the stars appears more or less constant, but the brightness of the planets varies widely on scales as short as weeks and as long as years.

In order to account for the complicated motion and varying brightness of the planets, Ptolemy described their motion in terms of smaller circles attached to the large circles that carry the planets around the earth. These smaller circles are called *epicycles*. As the planets move around their epicycles, they appear to move forward and backward against the background of the stars. They also move closer to and farther from the earth and in the process appear brighter and then dimmer.

Ptolemy is not in great repute now, so it is easy to forget that a description of the world that endured for 14 centuries is no mean feat. Ptolemy's model was sophisticated, and he never pretended to describe how the solar system actually works. Rather, he made quite clear in his great work, the *Almagest*, that he was presenting a *model* to allow the positions of the planets to be calculated, not a description reflecting the way the planets actually move.

When we think of a model, we usually imagine a miniature that captures at least something of the structure of whatever the model represents. Ptolemy was using the notion in a somewhat different and, as we shall see, very modern way. Ptolemy's model is a device for calculating the position of the planets, not a device for representing the appearance of the planets as they would look to anyone who had the entire system in view.

People have said that Copernicus's description of the solar system was both simpler and more accurate than Ptolemy's. In fact, Copernicus's description is neither. Some commentators have said that Ptolemy's model required the addition of more and more epicycles to continue to describe the motion of the planets as the centuries went by, but such embellishments were not needed. As far as complexity is concerned, Copernicus's model makes use of epicycles, just as Ptolemy's does. Copernicus's calculations lead to more accurate predictions than Ptolemy's in some instances, less accurate predictions in others. Why then is Copernicus's view so attractive?

In principle, the Copernican system simplified the motions of the planets by explaining some of their strange behavior as a reflection of the earth's motion. The appearance that the planets change the direction of their motion against the background of stars is not a "real" motion but an apparent motion, similar to the illusion that the stationary train next to yours is moving backward when yours in fact is moving forward. In practice, Copernicus still needed epicycles to describe the motion of the planets, although he got by with smaller ones than Ptolemy needed.

Copernicus was willing to overlook the complexities of the model he was forced to construct to describe the motion of the planets because he found the concept of a sun-centered solar system so much more attractive aesthetically than the Ptolemaic system:

> In the middle of all this sits the Sun enthroned. In this most beautiful temple could we place this luminary in any better position from which he can illuminate the whole at once? He is rightly called the Lamp, the Mind, the Ruler of the universe. . . . So we find underlying this ordination an admirable symmetry in the Universe, and a clear bond of harmony in the motion and magnitude of the Spheres.[3]

Scientists are often attracted by the aesthetic simplicity and symmetry of a theory. Einstein was particularly so. He was so sure his theory of general relativity was correct that when asked what he might have felt had the observations failed to confirm the theory, he responded, "Then I should have been sorry for the dear Lord—the theory is correct."[4] A devotion to beauty and simplicity can be very motivating but also badly misleading.

The Ptolemaic system fits nicely with Aristotle's physics. Aristotle was a student of Plato's and, like many early Greek natural philosophers, was not always concerned with making detailed observations. This disregard for the details of the workings of the world is sometimes given as the major reason the Greeks did not develop science any further than they did. Aristotle was a participant in the ancient conversation based on the very attractive idea that we can uncover the truth by pure reasoning. The most dramatic example of the success of this technique is Euclid's *Elements*, in which he laid out the foundations of geometry and showed how to derive conclusions from these fundamental principles or axioms. Relying on logic alone works very well for mathematics, but the approach has serious shortcomings when it comes to describing the physical world.

Aristotle wondered why objects move the way they do—why rocks fall and flames leap upward. He decided that objects must have natural realms and that in their movements they are attracted to these natural realms. Heavy objects fall to earth because that is their nature. Light objects rise because that is their nature. For Aristotle, the earth is a realm of heaviness, change, and decay; the heavens are a realm of lightness and unchanging essences. The characteristic motion of heavenly bodies is circular because circularity is a constant and unchanging motion, unlike the motion of earthly objects, which fall until they reach the ground and then stop.

Aristotle's approach is embedded in a conversation focusing on the question *why*. We can always keep asking why (Why do objects fall according to their nature?) until no further answer emerges, but at least an explanation is initially attempted. As science developed, its practitioners became less and less concerned with this kind of explanation.

Copernicus replaced Ptolemy's model with a sun-centered model, but he made no attempt to replace Aristotle's physics. Copernicus thus had no explanation for why the solid and heavy earth is able to move in the same way as the heavenly bodies move, or why this motion cannot be sensed. Still, the sun-centered picture was so attractive that this shortcoming did not seem to bother Copernicus overly. Nor did it bother many of his followers, including the German astronomer Johannes Kepler at the turn of the seventeenth century.

Copernicus still spoke largely as the ancients did, but Kepler played a central role in creating the distinctive conversation that was to become modern science. Working at the beginning of the seventeenth century, Kepler was sure Copernicus was right; the sun must be the center of the solar system. For many years Kepler also firmly believed, along with Copernicus and the early Greeks, that the orbits of the planets must be circles or combinations of circles because a circle represents an ideal form. He never abandoned the early Greek Pythagorean notion that the spacing of the orbits of the planets is determined by the characteristics of the so-called regular solids—the tetrahedron, cube, octahedron, dodecahedron, and icosahedron.

Kepler departed from the tradition in the importance he placed on observing what the world was doing instead of figuring out what the world must be doing. Kepler was not an observer himself; however, during the closing decades of the sixteenth century the Danish astronomer Tycho Brahe had accumulated observations of the planets made with the precision instruments he developed. Although these observations were made before the invention of the telescope, Tycho's sharp eye and attention to detail produced a set of observations far superior to those Ptolemy or Copernicus had to work with.

Tycho had his own model of the solar system, intermediate between those of Ptolemy and Copernicus—a model Tycho was sure his observations would support. In Tycho's system the sun moves around the earth as it does in Ptolemy's system, but, as in Copernicus's system, the planets (earth not being one of them in this scheme) move around the sun. While this system seems more complex than either of its forebears, it has the interesting quality of being in agreement with all the observations that would soon support the Copernican system; we will see why shortly.

Kepler started his analysis by trying to match Tycho's observations of Mars to a series of circular orbits with centers offset from the sun. He finally achieved an orbit that eliminated the need for epicycles and predicted positions of the planet that differed by, at most, only 8 minutes of arc from Tycho's observations. Eight minutes of arc is approximately one-fourth the apparent diameter of the moon—not a very large discrepancy before the invention of the telescope. Kepler's model was considerably more accurate than Ptolemy's and Copernicus's, and were it not for Kepler's conviction that the ultimate test of a model is its agreement with observations, there the matter would have ended.

Kepler, however, was sure Tycho's observations were sufficiently accurate that a model that failed to reproduce his observations by as much as 8 minutes of arc just would not do. After years of work, Kepler reluctantly abandoned the idea that the orbits of the planets must be circular. He simply could not get circular orbits to fit Tycho's observations.

Kepler tried other shapes for the orbit of Mars, eventually settling on an ellipse—a "squashed" circle. An ellipse has two "centers," or focal points, while a circle has only one. By placing the sun at one of these focal points, Kepler could match the observations within the errors of Tycho's measurements. He found that similar elliptical orbits fit Tycho's observations of the other planets, but Mars is critical because Mars *requires* an elliptical orbit to fit Tycho's observations, whereas the orbits of the other planets are so close to being circular that they *could* be described by circular orbits. Kepler had at last discovered a way to simply and accurately describe the orbits of the planets: The planets move in elliptical paths about the sun with the sun at one focus. Gone were the epicycles of Ptolemy and Copernicus. The complex motions of the planets were now explained only in terms of the motions of both the earth and the planets as they move in elliptical paths around the sun.

The idea that the planets travel in elliptical orbits is called Kepler's first law. The idea that regularities in nature display laws at work is clearly an analogy drawn from human affairs. Of course, in the sixteenth century it looked quite the other way around—the laws of kings were seen as the earthly analog of the laws of God. For Kepler and his seventeenth- and eighteenth-century successors, the laws of nature were the divine instructions directing the behavior of the world. To uncover them was to discern God's blueprint for the universe.

But what explained the motion of the earth and planets? What replaced Aristotle's explanation? Kepler represented a step toward modern science in that he thought a force was needed to explain the motion of the planets around the sun—he envisioned something like a magnetic force drawing the planets around the sun. The American physicist and Nobel laureate Richard Feynman described a model similar to Kepler's in the following way:

> In those days, one of the theories proposed was that the planets went around because behind them were invisible angels beating their wings and driving the planets forward. You will see that this theory is now modified! It turns out that in order to keep the planets going around, the invisible angels must fly in a different direction and they have no wings. Otherwise, it is a somewhat similar theory![5]

When we look at Newton's contribution, we will see the direction in which these new wingless angels are said to fly.

Even before he drew the conclusion that the earth and planets move in elliptical orbits, Kepler had found a relationship between a planet's speed and its distance from the sun: A planet moves so the imaginary line connecting the planet with the sun sweeps over equal areas in equal periods of time; the closer the planet to the sun, the shorter the line and the faster the planet must move. This relationship is embodied in Kepler's second law of planetary motion. Kepler's third law describes the relationship between the distance from a planet to the sun and the time the planet takes to complete a circuit of the sun. These three expressions profoundly altered the conversation of physical science. They created a shift away from explanation (the question *why*) toward description (the question *how*) that was to continue for the next 300 years.

Ptolemy and Copernicus chose the epicycles in their models simply because they matched the observations. Kepler's orbits are much less arbitrary. Once the distance of a planet from the sun is given, for example, Kepler was not free to say it takes any time he liked to circle the sun; its period is fixed by his third law. When Galileo discovered the moons of Jupiter, Kepler's laws were found to describe their motions as well. Many years later the same laws were found to describe the motions of multiple star systems. Kepler went far beyond simply summarizing the description of planetary motions; he invented a way of talking about the motion of heavenly bodies that is still valuable today.

Description versus Explanation

The Italian physicist Galileo Galilei was a contemporary of Kepler's, well known for his troubles with the Church over the Copernican theory. The Church fathers were convinced that the Copernican worldview conflicted with the teachings of the Bible (after all, Joshua told the sun, not the earth, to stand still). Galileo was equally convinced that Copernicus was right, that the earth moves around the sun rather than the sun around the earth. At first, the Church fathers took a relaxed approach to Galileo's apostasy. They were willing to allow Galileo to teach the Copernican system as a way of *computing* the motions of the planets as long as Galileo did not teach that the planets *actually* move in the way the Copernican system describes.

After all, Ptolemy presented his system as a way of calculating the position of the planets, not as an explanation of how the system really worked. So the Church fathers were really asking Galileo to approach the question in the same fashion Ptolemy had. Galileo, however, was not noted for his inclination to compromise. He was persuaded that Copernicus had not only the better description of nature but also the *right* description. The Church disagreed and in those days the Church did not have to brook disagreement from anyone. In 1633, the Inquisition forced Galileo to renounce the Copernican view.

Galileo's refusal to temper his viewpoint is often held up as an example of scientific integrity in the face of religious dogma, but the situation was not quite that simple. First, the idea that we cannot tell which of two bodies is in motion is critically important to Copernicus's model—we cannot tell from the apparent motion of the sun whether the sun or the earth is moving. Second, Galileo himself developed arguments demonstrating that we cannot tell whether something is in uniform motion or at rest on the basis of observations made within the system. This principle is familiar to anyone who has ever traveled in an airplane moving at over 500 miles per hour. In this situation, an object dropped from a tray falls to the floor in exactly the same way the object would fall if the plane were motionless on the ground. As long as the air is smooth and the plane is not turning, climbing, or descending, we cannot tell, without looking outside, whether the plane is still or in motion.

Against these arguments, Galileo's intransigence seems courageous but not completely rational. Since Galileo's time, physicists have been increasingly successful in avoiding fights with religious leaders about who is right, although the current argument between biologists and religious fundamentalists about evolution recalls the conflict between Galileo and the Church.

Many years before his encounter with the Church, Galileo's enthusiasm for the Copernican system was fanned by the observations he made with the newly developed telescope. One of the first things Galileo observed was the rugged surface of the moon, which conflicted with Aristotle's view of the smooth perfection of the heavenly bodies. Galileo also discovered that Venus, like the other planets and unlike the stars, shows a disk. This disk changes in size and displays varying appearances just as the moon does, which indicates that Venus, like the moon, shines by reflected light. The changing appearance of Venus also supports the Copernican view. Venus could not display both crescent and full phases unless it sometimes passes between the earth and the sun and sometimes passes behind the sun—something the planet never does in the Ptolemaic system.

Tycho's model also calls for Venus to show phases. In fact, from the viewpoint of describing the motions of the planets, there is no difference between Tycho's approach and Copernicus's; both are equally accurate and predict the same appearances. In fact, Tycho's model is simply Copernicus's model viewed from the perspective of the earth—an operation modern physicists call a coordinate transformation, which is essentially a mathematical adjustment without physical consequences. Then why should one point of view be any more important than another? The answer is, it shouldn't. So why do we say that Copernicus and not Tycho was right? We will get back to this question when we look at Newton's contributions.

Galileo discovered four bright stars accompanying Jupiter and changing position with respect to the planet. He interpreted these observations as showing that there are satellites circling Jupiter, just as the moon circles the earth in the models of both Ptolemy and Copernicus. Galileo argued that Jupiter's moons are indirect evidence in support of Copernicus's system because they show heavenly bodies can move in paths around objects other than the earth. Since Jupiter is able to move and carry its satellites along with it, the earth could probably do the same thing, even though Galileo proposed no mechanism to explain how this behavior is possible.

Despite his modern viewpoint in many regards, Galileo was convinced that circular motion was fundamental and refused to pay attention to Kepler's demonstration that the planets move in elliptical orbits. Galileo is often regarded as the father of modern experimental science. But in his disregard of the observations on which Kepler based his model, Galileo showed himself to be more of a theorist than an experimenter. Galileo often seemed to use experiments to confirm conclusions he had arrived at by thinking about a problem. Much of his genius resided in posing the right questions—questions he was able to answer!

Galileo focused his attention on the question of how falling bodies behave. Aristotle maintained that the heavier an object is, the faster it falls to earth. Galileo's famous, although possibly apocryphal, experiment of dropping two objects from the leaning tower of Pisa and observing them strike the ground at almost the same time was a dramatic, if by then not wholly original, demonstration of the shortcomings of Aristotle's views. Galileo *showed* that all sufficiently massive objects fall with nearly the same speed. He *concluded* that, under ideal conditions, all objects fall with the same speed. The notion of ideal conditions became a very important aspect of the way physicists talk about the world. The question Galileo then asked was not *why*, which Aristotle might have asked—but what is the speed at which bodies fall and how does this speed change with time? By pursuing the question in this way, Galileo created the conversational shift from *why* to the *how much, how far,* and *how long* that characterize modern science.

Galileo rolled balls down ramps at many different angles; in the process, he convinced himself that under ideal conditions an object will fall four times as far in twice the amount of time: The distance a body will fall depends on the square of the time. These experiments also provided Galileo with a way to measure how far an object falls in each second. He concluded that the distance a falling object moves increases each second in the following pattern: 1, 4, 9, 16, 25, ... This conjecture allowed him to talk about falling bodies in a mathematical way that would have pleased the ancient Greek Pythagoreans as much as it did Galileo.

Galileo articulated what would become the guiding metaphor of physical science:

> The Universe, which stands continually open to our gaze, cannot be understood unless one first learns to comprehend the language and read the letters in which it is composed. It is written in the language of mathematics. . . .[6]

For Galileo the predictability of nature implied the existence of a language that science could learn to read. Aristotle attempted to read the book of nature, but in Galileo's view, Aristotle was not successful. As the English astronomer Sir James Jeans would say 400 years later, "From the intrinsic evidence of his creation, the Great Architect of the Universe begins to appear as a pure mathematician."[7] Whether or not God is a mathematician, we know that some human beings are mathematicians, and we will see that the language *they* speak will be interpreted more and more as being the language in which nature is written.

2

The Invaluable
Concept of Force

*Odd as it may seem, most people's views about motion are part of
a system of physics that was proposed more than 2,000 years ago
and was experimentally shown to be inadequate at least 1,400
years ago.*[8]

I. BERNARD COHEN

The Idea of Force

Imagine holding your hand out in front of you and tossing a ball upward. When the ball has traveled a short distance from your hand, is there a force acting on it? If there is, in what direction is the force acting? When the ball has reached the highest point in its path, just before it begins to fall back toward your hand, is there any force acting on the ball? Is there a force acting on the ball just before you catch it? If you are like the majority of us, including many who have taken a course in physics, the answers you give to these questions will not agree with the answers physicists give. This difference is what Cohen referred to in the quotation above. We will see how physicists developed the answers they give to these questions and why these answers differ from the answers most people give.

Kepler and Galileo developed descriptions of motion that are useful because they can be used to make predictions. Isaac Newton, however, carried the process much further than any of his predecessors. In his *Principia*, published in 1687, Newton outlined a series of fundamental principles, or laws, and used these to calculate how objects move in a wide variety of circumstances. In the process he developed the science of mechanics in essentially its present form.

Newton invented a way of talking about motion that paralleled Euclid's way of talking about geometry. Newton invented a series of axioms, or laws, of motion. From these axioms he was able to generate theorems, and these theorems could be interpreted, among other ways, as descriptions of the motion of the planets.

In order to discover the consequences of the principles he formulated, Newton was forced to develop a new language—the language of what he called *fluxions* and we call *differential calculus.* Before Newton, there was no way to describe quantitatively the changes brought about by motion—mathematics could be used to talk only about static or unchanging situations. Newton described motion as continuous. Between any two points on the path of an object, he said, there is always another point. The calculus allows such continuous motion to be described quantitatively. When physicists followed Newton's approach, the calculus became the language of physics.[9]

Newton began by drawing a distinction between motion that requires explanation and motion that does not. For Aristotle, *all* motion must have a cause. Galileo imagined an experiment that convinced him horizontal motion on a frictionless surface would continue indefinitely—once an object in this situation is put into motion no force is required to keep it in motion. In other words, for Galileo, uniform motion in a straight line requires no explanation. Indeed, we do not get anywhere asking why a body in uniform motion remains in motion; according to Galileo, this is simply the way the world is.

Even while talking about imaginary frictionless planes, Galileo remained reasonably close to examples from the physical world, but Newton broke this connection. He generalized Galileo's way of speaking by declaring that *all* uniform motion will continue indefinitely unless something interferes with it—the principle of *inertia.* The only behavior that needs explanation, in Newton's view, is motion deviating from a straight line or motion that speeds up or slows down.

Now we have lots of experience telling us that a body in motion *does not* remain in motion. A physicist explains that friction normally brings moving bodies to a stop, but when friction can be minimized, we can see that Newton was right: Objects *do* continue in uniform motion. Nevertheless, we have no experience of such a Newtonian world; Newton's laws describe an ideal world, not the one we encounter every day. Working with ideal arrangements that minimize complications made possible great advances in physics, but this approach took physicists further and further from the world of experience into a world of abstraction.

Kepler gave up the idea that circular motion in the heavens is natural; he came to believe that a force is needed to move the planets around the sun. On the other hand, for Galileo, the motion of the planets in their orbits around the sun needed no merchanism—circular motion was as natural for Galileo as motion in a straight line. Newton followed Galileo in saying that a continuing force is not needed to maintain motion in a straight line. Unlike Galileo, however, Newton required a mechanism to maintain the planets' motion on *curved* paths.

As presented in the *Principia*, Newton's first law of motion is as follows:

Every body perseveres in its state of being at rest or of moving uniformly straight forward, except insofar as it is compelled to change its state by forces impressed upon it.

The first law implies that if an object deviates from uniform motion, something is causing it to deviate. But what sort of thing? And how will the object respond?

Newton's second law answers some of these questions. It declares:

A change in motion is proportional to the motive force impressed and takes place in the direction of the straight line along which the force is impressed.

If a force is exerted on a body it will produce a change, either in the direction of motion or the speed of the body, and the change will be in the same direction as the force is applied. The second law also allows us to calculate the amount of change. Newton did not tell us what a force is, but he did give us an example—gravity.

The story of the apple, whether true or not, contains the essence of Newton's bold extrapolation that the *same* language can be used to describe the way objects behave on the earth and in the heavens. Newton described gravity as a force acting between any two bodies in the universe, a force depending on the mass of the bodies and growing weaker as the square of the distance between the bodies grows greater. The dramatic demonstration of the power of this description was Newton's calculation showing that the moon "falls" each second toward the earth exactly the distance it would be expected to fall if the same force controlled its motion as controls the fall of an apple in an orchard, once account is taken of the greater distance to the moon. The moon does not hit the earth, as the apple does, because part of its velocity, not provided by gravity, is at right angles to the line connecting the moon and the earth. This part is sufficiently great that the moon wants to escape the earth but is prevented from doing so by gravity—the moon effectively "falls" around the earth. When Feynman said the wingless angels that move the heavenly bodies face in a different direction, he meant that instead of pushing the moon around in its orbit, the angels push it toward the earth. Clipping the angels' wings was the inevitable simplification that mathematics brings.

Now we can return to the ball discussed at the beginning of this chapter. For Aristotle, and apparently for many of us, there is a force acting on the ball, moving the ball upward after it leaves your hand; at the height of the ball's path, when it is motionless, that force has been expended. The ball then returns to the earth, its natural dwelling place. When the ball begins to fall, those of us who think in terms of gravity, as Aristotle did not, say gravity is the force pulling the ball toward the earth. Otherwise, our explanation may not differ very much from Aristotle's.

In the language of Galileo and Newton, there is no upward-directed force acting on the ball after it leaves our hand. The ball moves upward because once set in motion it has the tendency to remain in motion. In Newton's language, what requires a mechanism is not the ball's upward movement, which was imparted by your hand, but the fact that the ball slows down. Here again Newton's language provides a description. The ball slows down as a result of a *force* slowing it down—gravity is pulling the ball back to earth. At the highest point of the ball's path, gravity has overcome the upward velocity imparted by your hand when you released the ball. At this point the ball is motionless, but for Newton the force is still acting and is directed downward. Just before you catch the ball, the force of gravity is still acting on the ball, accelerating it toward the ground, just as it has been all along. When you catch the ball the force of gravity is still acting on the ball, but you are exerting a force that exactly balances the force of gravity, and so the ball remains motionless.

In the Newtonian picture, at all points in the ball's path the same force is acting on the ball, and the force always acts in the same direction. The predictive power of Newton's language is so great that physicists adopted it without ever returning to Aristotle's language.

It is not simply that Newton's way of talking about force is right and Aristotle's way wrong. In a world where friction is as important as it is in our everyday life, Aristotle's language fits our experience reasonably well—objects *do* require a force to keep them in motion. After all, if so many of us, including physicians, mechanics, and baseball players, continue to think of force in the same way Aristotle did, this approach cannot be all bad. The superiority of Newton's approach to Aristotle's emerges clearly only when someone is concerned with describing motion quantitatively.

Newton was able to show that Kepler's laws follow from Newton's laws of motion and his mathematical expression for gravitational force. According to Newton, Kepler's second law follows from his own laws if the planets move in response to a force directed toward the sun. Newton also showed that the motion of the planets in closed orbits—the ellipses described in Kepler's first law—follows if there is a force controlling their motion that grows weaker with the square of the distance, the kind of behavior that Newton associated with gravity. Finally, Newton demonstrated that Kepler's third law, relating the distances of the planets from the sun to the time the planets take to circle the sun, also follows from the way the attractive force of the sun's gravity weakens with distance.

Once Newton's language is adopted, the motions of the planets follow—the earth moves in a way that is consonant with the laws of motion. We do not observe the earth circling the sun; this motion is an interpretation made in light of a theory. As seen from the earth, both the sun and the moon appear to circle the earth roughly once a day, and the sun traces a path around the background of stars once a year. From the surface of the sun, the earth would seem to circle the sky once every 26 days, because the sun takes 26 days to turn once on its axis. Someone on the sun would see the earth tracing a path against the background of stars once a year, just as the sun seems to do when viewed from the earth. If we fixed ourselves at a point above the north pole of the sun and did not rotate with the sun, the earth would still appear to circle the sun, but the moon would also appear to circle the sun, weaving a path around the earth. The model we have is just that—a model. Thanks to Newton, this model works very well in predicting the motions of the planets. Tycho's description is no longer in vogue because no dynamic theory supports it—no model plausibly describes how all the planets except the earth circle the sun, while the sun circles the earth.

The way of speaking Newton created could be applied to a wide variety of circumstances ranging from falling apples to spacewalking astronauts. Furthermore, Newton showed that a limited number of principles can be used to describe a wide variety of phenomena. The success of Newton's language was so dramatic that it seemed to be a conclusive demonstration of the fundamentally mechanical character of nature.

Newton, like most of the natural philosophers of the seventeenth and eighteenth centuries, was religious. In fact, his theological writings, although largely unpublished, are as voluminous as his scientific writings. Newton saw himself as uncovering God's design—the laws God used in fashioning the universe. In Newton's view, God said, "Let objects in motion remain in motion." Despite this piety, however, the very success of Newton's language began a process of undermining the theology he valued so highly.

Newton believed not only that God had created the universe and set it in motion but also that He continues to guide it and make the corrections necessary to keep it on track. Others, however, began to wonder if such corrections were really necessary. And if God's only function was to start the universe in motion, perhaps He was not as indispensable as Newton thought. When the French mathematician Pierre Laplace gave a copy of his magnum opus, *Mécanique céleste,* to Napoleon, the emperor is reported to have asked him if it was true that this book on the structure of the heavens never even mentioned the Creator. Laplace is said to have responded, "I have no need of that hypothesis." Once Kepler initiated the process of making observation the ultimate authority for physical descriptions, the success of science seemed to call for making observation the *only* authority for *all* descriptions.

Despite their remarkable success, Newton's formulations raised a number of questions. Some people rebelled against the mechanical nature of Newton's approach. The late eighteenth- and early nineteenth-century romantic poets saw the success of the Newtonian approach as a blight rather than a blessing. In Wordsworth's words:

Sweet is the lore which Nature brings;
Our meddling intellect
Mis-shapes the beauteous form of things:-
We murder to dissect.
Enough of Science . . . [10]

Or William Blake's:

May God us keep
From Single vision & Newton's Sleep![11]

Even those who did not object to the fundamentally mechanical character of Newton's approach had questions. For example, some were troubled that Newton's approach was not mechanical enough! They were bothered by the idea that gravity acts over vast empty spaces. And the idea that the sun controls the motion of the distant planets without any intervening mechanism seemed puzzling at best.

Some aspects of the laws of motion invented by Newton preserve a connection with experience. For example, the idea that a force is necessary to produce an acceleration fits with our experience that we must exert ourselves to move a heavy object. On the other hand, we have no direct experience of a force of gravity nor of anything pulling us toward the earth as we stand or walk. The idea of forces acting at a distance with no intervening mechanism is not based on experience; such forces are fundamentally mathematical in nature. No verbal explanation seems to make Newton's expression for the force of gravity more intuitive, and none is needed to allow anyone with the necessary mathematical skills to use the expression effectively.

In the face of questions about *why* gravity works, Newton responded, "I do not make hypotheses." By this he meant he did not attempt to tell us *why* gravity works the way it does. Instead he told us *how* gravity works—how to predict the *effects* of gravity. Although Newton's description of the world is mechanical in that it involves forces and the actions produced by forces, it is not mechanical in that he offers no explanation for the *mechanism* of gravity. Instead, he gives a mathematical expression that allows someone to calculate the effects of the gravitational force at any point in space. In the tradition of Galileo, Newton spoke about *how* something occurs rather than *why* it occurs.

Newton created a way of talking about nature based on a declaration that uniform motion requires no explanation. The first law of motion is aptly described, because like any law, it cannot be proved. Clearly there is no evidence that things will continue to move forever in a straight line. We have not been around long enough to justify such a statement, nor do physicists have an experimental arrangement that they could use to test the assertion if they wanted to. Newton's assertion is not a conclusion; it is an assumption. The value of Newton's first law is not that by itself it can be proved or disproved. In order to be useful, the first law must be considered along with his other laws. This combination, together with a description of the bodies on which the forces are acting, allows predictions to be made of the motion of the bodies. Aristotle's vocabulary has disappeared from the language of physicists because it does not lead to quantitative predictions, and testing predictions is the only process physicists recognize as valid in determining the value of a way of describing nature.

The Concept of Energy

Following Newton, work on mechanics and the calculus was pursued by a number of creative mathematicians and physicists. In the process they invented a way of speaking about motion that does not involve the language of force at all. Instead, they took a fundamentally different approach to describing the behavior of a mechanical system, relying on the idea of *energy*. To understand the significance of this approach we must look a little more closely at energy.

Imagine a roller coaster in which the cars are suspended by air cushions to minimize friction. After the cars are pulled to the top of the first hill, the only force acting on them is gravity; unlike automobiles, they have no motors to keep them in motion. It is not hard to believe that a roller coaster has energy as it rushes down a hill and reaches the bottom. If we stop the cars suddenly (as sometimes occurs inadvertently on the highway), we find a lot of twisted and bent metal. Something had to do all that bending and twisting, and we call that something energy. But exactly what sort of a something is energy? When physicists have questions, the first thing they look for is something to measure. How fast the cars are moving seems like a good place to start.

The speed of the cars does not turn out to be a good measure of their energy, although physicists took a while to recognize this. In the language of mechanics, the speed of the cars is related to something physicists call *momentum*. The energy is related to the square of the car's speed (the speed multiplied by itself). This relationship is the reason an automobile accident at 60 miles per hour is much worse than one at 30 miles per hour (four times as violent as far as the energy is concerned). If the roller-coaster cars have all this energy at the bottom of a hill, where does the energy go when the car climbs to the top of the next hill? Does it get used up in the process of pushing the car up the hill? If so, how does the energy get regenerated on the way down the other side of the hill?

Instead of answering these questions directly, physicists draw our attention to something interesting. If we measure the speed of the cars, we can use this speed to develop a number that physicists call the *kinetic energy* because it involves motion. Physicists use the height of the cars above the ground as part of another number, which they call the *potential energy* because it describes a potential that can become actual as the cars roll down the hill and gather speed and, hence, kinetic energy. Almost always these two numbers are different, but if we add the two numbers we get a third number: the *total energy*. For any point on the roller-coaster track, this number is *always* the same! This surprising result means that for any point on the roller coaster's route, we can measure the height of the track from the ground and figure out how fast the cars will be moving when they reach the point. Furthermore, this approach works no matter what the shape of the path of the roller coaster. Why does it work, and what does its success mean?

Physicists describe the calculations by saying that the energy of the roller-coaster cars is *conserved* (meaning that the third number, the total energy, does not change). Furthermore, they say that the cars exchange kinetic energy and potential energy, by which they mean that when the kinetic energy is high, the potential energy is low (the cars at the bottom of a hill move very rapidly); when the kinetic energy is low, the potential energy is high (the cars at the top of a hill move very slowly).

Physicists have built what they call a *model* of the roller-coaster system. The model is a series of equations (simple equations in this case) that can be used to predict the behavior of the cars. It focuses on two aspects of the cars—their height above the ground and the speed at which they move—and relates numbers associated with these aspects in such a way that knowing the total energy and either the height or the speed, we can calculate the other for *any* roller coaster.

While physicists can describe their calculations in words, they do not explain what lies behind the calculations. For example, they have no idea *why* the cars exchange kinetic energy and potential energy. Furthermore, they have no idea *where* the kinetic energy goes as the cars climb the hill. Instead, they have found some very useful numbers and described how to calculate those numbers. Physics is less about explanations than about knowing how to calculate useful numbers.

The roller coaster is not totally frictionless. If we made very careful measurements we would discover that the sum of the kinetic and potential energy of the cars is not always exactly the same but slowly decreases in time—the cars cannot reach the same height after they have been traveling for a while. Does this mean energy is *not* conserved? Physicists say no. They say if we measure the temperature of the cars, the tracks, and the air, we would find that they are slowly getting warmer. If we added this increase in *thermal energy* to the potential and kinetic energy of the cars, we would find *this* number remains constant. Energy, they assure us, is still conserved.

"Wait a minute," the suspicious are likely to say. "Physicists called the sum of the kinetic and the potential parts the total energy and said this number does not change. When they found that it did change, they threw in another number that they called thermal energy to make the books balance. An accountant who played this fast and loose would probably wind up in jail, so how can physicists get away with this kind of bookkeeping?" A physicist might offer the following defense: "Nothing *guarantees* that I can find a number that remains the same no matter how long the cars have traveled, how high they are on the tracks, or how fast they are moving. It is one thing to simply state that something is a source of energy; it is quite another to demonstrate that a number has precisely the *magnitude* necessary to ensure the conservation of energy. Furthermore, these formulas apply to all sorts of different systems besides roller coasters—pendulums and planetary orbits, for example. This success gives me confidence that what I am talking about is real and not just a bookkeeping trick."

Physicists continually ignore the connections between the system they are studying and the rest of the world, including themselves. They also often ignore major characteristics of the systems; the value of this procedure is demonstrated by its success. In discussing the roller coaster, physicists are not concerned with the shape of the track or the materials used to construct either the track or the cars because the question they are interested in answering is how fast are the cars moving? Once they know the total energy, they can derive the speed of the cars from their height without having to know any of the details of the composition of the cars or their precise path. The height and speed are related in such a way that physicists can ignore all other details, simplifying things immensely.

Talking in terms of the conservation of energy has an additional benefit for physicists: It allows them to rule out an unlimited number of "forbidden roller coasters" (forbidden because the design would violate the conservation of energy). A simple example of a forbidden roller coaster would be one topping a second hill that is higher than the first. Let us see why this will not work. At the top of the first hill the cars are not yet moving (they have been pulled up by a motor attached to the fixed structure and ideally can come to a complete stop before starting down again). This means the kinetic energy of the cars is zero, and so the total energy is equal to the potential energy. From this point on the cars move under the influence of gravity alone, without any additional help from a motor. If the second hill were higher than the first, at the top of the second hill the cars would have even more potential energy than they did at the top of the first. But for the total energy to remain constant, which the conservation of energy demands, the kinetic energy of the cars at the top of the second hill would have to be *less* than zero. (If the potential energy is larger, the kinetic energy must be smaller—but the kinetic energy was zero to begin with.) Since we cannot imagine a car moving at a speed of less than zero miles per hour, we know that such a roller coaster would not work. Describing the world in terms of what can happen and what is forbidden from happening will play an increasingly important role in the way twentieth-century physicists speak.

Hamilton's Method

At the end of the eighteenth century, the Italian-French mathematician Joseph Lagrange showed that in talking about any system where friction can be ignored, it is possible to define a potential, similar to the height of the cars in our roller coaster. He also showed that for any mechanical system, he could define a kinetic energy similar to the energy of the cars in our roller coaster. Lagrange formulated a mathematical expression, called the *Lagrangian* in his honor, which is simply the difference between the kinetic and potential energy for every point in an object's path.

In 1824 the Irish mathematician William Hamilton declared that an object or collection of objects moves in such a way that the *action*—a quantity closely related to the Lagrangian, or the energy of the system—always has the smallest possible value. Hamilton started by adding up the differences between the kinetic and potential energies for every point along the actual path taken by a system. Then he carried out this procedure for a number of imagined routes that would have allowed the object to arrive at the same destination in the same time. When he was through, he showed that the action for the route actually taken is always smaller than the action for any other route. In other words, a system moves in such a way that it minimizes the action.

Hamilton's way of talking about nature is rather amazing. For one thing, it does not give us the slightest clue about what is going on with the system in the way Newton's laws seem to. Hamilton's principle almost seems to imply that a system *knows* the various routes by which it can move from one location to another and *chooses* the route that gives the lowest sum for the differences between the kinetic and potential energies. Does this make any sense? Just as with Newton's force acting over distance, no reasonable explanation is forthcoming, no answer to the question *why*. Most physicists are content to say the system moves *as if* it knew the path that makes the action a minimum and let the question go at that. Here, as with Newton's refusal to speculate on why gravity works as it does, we see a mathematical formulation with predictive power replacing a tangible mechanical explanation.

Hamilton's approach can be used to develop a mechanics that is *not* based on Newton's laws. Whereas Newton's approach is formulated in the language of forces, Hamilton's adopts the language of energy. Which is the more fundamental of these two languages?

As far as a physicist is concerned, this question can be answered only by answering another question: How do the predictions of the two approaches differ in any situation? Despite their differences, the two formulations turn out to lead to exactly the same predictions. The reason for this apparent coincidence is simple: Hamilton's principle can be expressed in terms of Newton's laws. Although the two theories *seem* to provide different explanations for motion, physicists say the two approaches are equivalent. This equivalence means physicists cannot appeal to the outcome of any experiment to differentiate between the two theories, since both predict the same experimental results. Physicists maintain that the important aspect of a physical theory is what the theory allows them to predict, not the particular expression of the theory. The predictions that can be made by a theory follow from the mathematical formulation of the theory, *not* from the way that mathematical formulation is explained.

Classical mechanics can be used to describe the majestic sweep of the universe in a few simple laws. In Newton's language, these laws are expressed in terms of forces and the effects of forces on material objects. By talking about gravity as a universal force, and by formulating a simple mathematical expression to describe the effects of this force, Newton was able to describe a universe that seemed to be completely predictable. The dream of the astrologers was fulfilled; the motions of the planets pointed the way to predicting the future. But the future was not written in the language of the stars—the future was written in the language of Newton.

3

The Indispensable Idea of Fields

A courageous scientific imagination was needed to realize that not the behavior of bodies, but the behavior of something between them, that is, the field, may be essential for ordering and understanding events.[12]

ALBERT EINSTEIN

If you go out on a clear dark night anywhere in the Northern Hemisphere and have some familiarity with the stars, it will not take you long to locate the fairly bright star that marks the pivotal point about which the sky seems to turn. Had you gone out on a similar night 10,000 years ago, that central point would be elsewhere in the sky, far from any conspicuous marker. Those who depend on the sky for their directions are fortunate that Polaris serves to mark north so clearly.

The stars guided sailors for millennia, but daylight and clouds made it very risky to venture too far from shore. In the twelfth century, European sailors began to employ a new technology, possibly brought back from China. A peculiar kind of rock known as lodestone ("leadstone") seemed to mysteriously "know" the direction of the polestar and point toward it, no matter what the weather or time of day. Needless to say, this technology revolutionized travel at sea, although for many centuries magnetism smacked of the occult.

An equally mysterious "force" had been known since the time of the early Greeks—the ability of amber rubbed with a piece of silk or fur to attract pieces of lint. *Electron* is the Greek word for amber. Although magnetism found immediate applications, *electricity,* was slower to find a practical role. It was not until the seventeenth century that people recorded that two glass rods each rubbed with silk repel each other, and it was not until 1800 that Alessandro Volta constructed the first voltaic cell or battery.

Electricity and Magnetism

It should come as no surprise that the success of Newton's language in describing gravity was a powerful incentive for others to attempt to apply the same vocabulary to the equally mysterious "forces" of electricity and magnetism. Electrical force differs from gravitational force in that there are two kinds, or *polarities,* of electricity, labeled, for no profound reason, *positive* and *negative.* Electric charges of opposite polarity attract each other the way two masses attract each other in Newtonian physics. Electrical charges of the same polarity, however, repel each other, which has no gravitational analogue. It is well to bear in mind that electricity is no more a "something" than anger is a "something"; electricity is a way of talking about how things behave.

Magnetism can be described in much the same language as electricity. One end of a magnetized bar will attract one end of a second magnetized bar and will repel the other end of the second bar. However, it turns out to be impossible to separate the two polarities of magnetism in the same way it is possible to separate electric charges. If we cut a magnet in half, we do not wind up with pieces of different magnetic polarities; instead we wind up with two smaller magnets, each with the same two polarities. Because magnets were used for centuries to find direction, magnetic polarities are referred to as north and south poles, rather than plus and minus charges.

Electrical and magnetic forces display a wider variety of effects than gravity does, and it was not until 200 years after Newton's *Principia* was published that a unified description of electricity of comparable power was developed. A hundred years after *Principia,* at the end of the eighteenth century, the French scientist Charles Coulomb showed that the interaction between stationary electric charges could be talked about in the same way that Newton talked about gravitational force—in terms of a force that weakens with the square of the distance as gravity does. In 1821 the Danish physicist Hans Oersted showed that when electricity flows through a wire, a nearby magnetic needle will move; electricity and magnetism seemed to be related in some way.

If the behavior Oersted observed is described in terms of a force exerted by the electricity, the force must be unlike gravity because it does not act directly on a line between the wire carrying the electric current and the magnetized needle but rather as if the force circled the wire.

The French physicist André Ampère wasted little time in using Oersted's discovery as the basis for a description of electrical force framed in Newtonian language. But it was the brilliant English experimental physicist Michael Faraday who used Oersted's find to develop a new vocabulary that has profoundly shaped the language of physics down to the present day.

In 1831, Faraday and the American physicist Joseph Henry independently discovered that a changing magnetic force can produce an electric current. Rather than talking about electric charges interacting over an intervening empty space, the way Newton talked of particles interacting gravitationally over large distances, Faraday described his experiments with electricity and magnetism in terms of a *field* surrounding electric charges and magnetic poles.

Think of two electric charges of opposite signs separated by some distance. Let us call them Plus and Minus. Plus and Minus are electrically attracted, just as two bodies are gravitationally attracted. In Newton's and Coulomb's description this attraction is in the form of a response at a distance—let us say Plus responds to Minus by moving toward the distant Minus, which is held stationary. For Faraday, Plus responds, not to the distant Minus, but to a field of influence produced by Minus and distributed throughout space. One illustration of Faraday's field is the familiar pattern traced by iron filings in the vicinity of a bar magnet.

Unlike Newtonian gravity, which depends solely on the masses of the objects involved, electrical and magnetic effects depend not only on the strength of the electrical charges and magnetic poles but on their motion as well. This complication was most easily incorporated into physical descriptions by making use of Faraday's field. Faraday's field proved to be related to a phenomenon that at first seems to have nothing to do with electricity and magnetism—light.

The Nature of Light

In addition to his remarkable accomplishments in describing the effects of gravity, Newton also made significant advances in describing the behavior of light. For example, he demonstrated that "white" light can be thought of as being made up of all the colors of the rainbow. This led Newton to solve one of the major problems of the telescopes of his day, the appearance of colors surrounding images of white objects.

Seventeenth- and eighteenth-century physicists were divided about the nature of light. Like the ancient Greeks, Newton talked about light as a series of tiny particles flowing out from a source. Although the majority of physicists accepted Newton's model, the Dutch scientist Christian Huygens devised an alternative one. He talked about light as a wavelike phenomenon, similar in many respects to the waves created when a pebble is thrown into a pond. Unlike the two approaches to describing motion put forward by Newton and Hamilton, which make the same predictions, the two ways of talking about light predict different effects and hence can be distinguished by measurements.

When a stick is submerged partially in water at some angle to the vertical, it seems to bend at the waterline in such a way that the part under the water appears more vertical than the part in the air. This observation can be explained by a simple model in which particles of light are attracted by a force at right angles to the surface of the water. In this description, light speeds up when it leaves the air and enters the water. In the wave description, on the other hand, if light slows down as it encounters the water, the wave will "pivot" to produce the same effect. The technology of the seventeenth and eighteenth centuries was not sufficiently developed to make measurements of the speed of light in two different media with the accuracy necessary to determine which way of talking was most consistent with observations.

Lack of decisive evidence, however, has never deterred people from taking positions, and physicists are no exception. Arguments were mustered on both sides. Newton, for example, supported the particle view. He reasoned that light must travel in straight lines. It is easy to imagine a series of tiny particles moving in straight lines outward from a light source. Waves, on the other hand, *can* bend around corners. This objection led Huygens and other supporters of the wave viewpoint to argue that the wavelength of light must be very short, since if wave crests are very close together the amount of bending will be very small.

In 1650 the French mathematician Pierre de Fermat articulated a principle describing the behavior of light. Fermat's approach resembles Hamilton's description of the motion of mechanical systems (or rather Hamilton's description resembles Fermat's, since Fermat was first), because it seems to call for an inexplicable behavior on the part of light. Fermat said light travels from one point to another in such a way that it takes the least amount of time to make the journey. Fermat's *principle of least time* successfully describes reflection and refraction and allows opticians to design lenses, but it gives them no clue about why there is a unique route, if there is one, or how light is able to choose this unique path from among all possible routes.

Experiments performed at the beginning of the nineteenth century by the English scientist Thomas Young and the French engineer Augustin Fresnel played important roles in resolving the dispute over the nature of light. They essentially demonstrated that light *is* able to bend around corners in the way we would expect if light were a wave. Young set up a demonstration in which light from a source fell on a screen with two slits. Behind the screen was a second screen on which the light passing through the slits produced an alternating pattern of light and dark bands parallel to the slits.

When two waves meet, the crests and troughs add to produce a new wave. The bands Young observed are interpreted as areas where the waves passing through the two slits are either in phase (crest matching crest) and add to reinforce each other, producing bright bands, or out of phase (crest matching trough) and cancel each other, producing dark bands. The particle model provides no such simple explanation for the bands. From our perspective, the language of Huygens (waves) emerged triumphant over the language of Newton (particles), but from the perspective of physicists in the latter half of the nineteenth century, Huygens was clearly right and Newton clearly wrong—light must be a wave and night a particle phenomenon. We will see that this judgment was premature.

Electromagnetic Radiation

In the mid-nineteenth century, a Scotsman, James Clerk Maxwell, developed a mathematical expression for Faraday's field as elegant as Newton's laws of motion. Maxwell's laws are more abstract than Newton's. In fact, they normally are not expressed in words at all but only as mathematical formulas. Nevertheless, Maxwell based his formulas on a mechanical model of the way in which electricity and magnetism work. This model envisioned all space filled with invisible "vortex tubes."

Maxwell was able to describe the electric and magnetic behavior known at that time, but he went far beyond what was known. Based on his mechanical model, Maxwell postulated the existence of an as-yet unobserved phenomenon—a magnetic field produced, not by an electric current as Oersted demonstrated, but by a changing electric field.

Maxwell's equations called for a changing electric field to produce a magnetic field and a changing magnetic field to produce an electric field. This mutual relationship leads to the following situation. A changing electric field produces a changing magnetic field in its immediate vicinity. Since this magnetic field is changing, it in turn produces a changing electric field in its vicinity, and so on. The result is a wave of changing electric and magnetic fields propagating through space at a fixed speed.

Although we talk about the speed at which an electromagnetic wave moves, it is important to keep in mind that in the wave description no physical object is going anywhere at this speed. Instead, at any particular point in space, the strengths of the electric and magnetic fields are changing. Imagine a source of radiation is turned on. The fields at a point some distance from the source will begin to oscillate only after a period of time that represents the time it takes for the wave to "reach" this point. Unlike gravity, which in Newton's model propagates at an infinite speed, electromagnetic effects travel at finite speed. In Maxwell's model, the speed with which electromagnetic radiation moves turns out to depend on the strengths of the electric and magnetic forces. The value Maxwell calculated for this speed closely matches the measured speed of light, a surprising agreement that led him to conclude that light must be an electromagnetic phenomenon.

If you have difficulty picturing electromagnetic waves, you are in good company. Richard Feynman said, "I have no picture of the electromagnetic field that is in any sense accurate. . . . It requires a much higher degree of imagination to understand the electromagnetic field than to understand invisible angels. . . ."[13] Nevertheless, the electromagnetic field would prove a very durable way of talking about nature.

At first, not a great deal of attention was paid to Maxwell's work. The reason was simple: There seemed to be no way to test Maxwell's predictions. There was no way of creating a vibration fast enough to produce light waves. Twenty years after Maxwell's publication, the German physicist Heinrich Hertz succeeded in producing electromagnetic radiation with wavelengths much longer than those of light—radio waves. He carried out a series of experiments demonstrating that these waves reflect, refract, and in general behave exactly as light does. Hertz's demonstrations convinced virtually all physicists that Maxwell's equations accurately describe the behavior of light.

Physicists envisioned that the electromagnetic field required a supporting mechanical structure that played a similar role to that played by water in an ocean. Only instead of water, physicists pictured a structure filling all of space, called the *luminiferous ether*. The idea of the ether, or aether, goes back to Aristotle, for whom it was the pure and incorruptible realm from which the heavens are constructed. The idea underwent changes over the following 2,000 years, and by the end of the nineteenth century the only function of the ether was to serve as a medium in which electromagnetic waves can propagate. The ether, however, had some very unusual properties.

The ether had to be very rigid to support the rapid changes associated with the passage of electromagnetic waves. But it offered no resistance at all to material bodies, such as the planets as they circle the sun. Despite this unusual combination of properties, the ether seemed essential to support, literally and figuratively, the electromagnetic field. The first evidence of a fundamental problem with thinking about the ether in this way emerged as the result of a series of measurements whose outcome at first seemed a foregone conclusion.

Does It Move, After All?

The ether provided the opportunity to demonstrate at last that Galileo's legendary sotto voce challenge to the Inquisition was correct. (After formally disavowing the Copernican heresy, Galileo was said to have murmured under his breath, "Nevertheless it moves.") Physicists set out to measure the motion of the earth as it passes through the ether. The result was disappointing, to say the least.

After repeated attempts, at the end of the nineteenth century the American physicists Albert Michelson and Edward Morley, despite using remarkably sensitive instruments, were unable to detect the motion of the earth through the ether. Was Galileo wrong and the Church right after all? The evidence that the earth moves around the sun is so overwhelming that, rather than giving up this way of talking, physicists chose to conclude, not that the Earth does not move, but that the ether does not exist. Numerous experiments, including those of the Italian Guglielmo Marconi, developer of the radio, seemed to demonstrate the existence of electromagnetic waves beyond any question. How could electromagnetic waves exist without a medium? How can we talk about waves in a pond and maintain at the same time that there is no water in the pond? Ultimately, the resolution of this conundrum required a virtual abandonment of the amazingly successful worldview of eighteenth- and nineteenth-century physics.

Physicists are pragmatic and were understandably unwilling to abandon the descriptive and predictive power of Maxwell's equations. Instead they accepted the idea that the model on which Maxwell based his theory was dispensable and took the position that Maxwell's mathematical expressions *are* his theory. In other words, physicists gave up the idea of a mechanical model, but they retained Maxwell's mathematics. Only the smile of the Cheshire cat remained.

Models of the Physical World

We have traced the evolution of ways of talking about the physical world through several stages. Aristotle explained motion in terms of the proclivity of earth, air, fire, and water to seek their own realm. This explanation had no *quantitative* dimension, however. Ptolemy did not attempt to describe the motions of the heavens so much as he provided a way of calculating these motions. Ptolemy's method prevailed for 14 centuries.

In the sixteenth century Copernicus argued that the sun, rather than the earth, must be at the center of the universe. Despite its appeal, Copernicus's heliocentric description with its epicycles and circular planetary orbits did not represent a substantial improvement over Ptolemy's approach to predicting the motion of the planets. Dramatic improvements in such predictions required Kepler's invention of three laws to describe planetary motion—a description in which the planets move about the sun in elliptic orbits. Finally, Newton developed a language that allowed him to deduce Kepler's laws from even more fundamental principles—descriptions that Newton said applied to all systems moving under the influence of gravity. Newton's approach is based on forces and their effects, but an alternative way of talking about motion in terms of energy proved equally powerful in describing the behavior of the world. Since both approaches, despite their apparent differences, lead to the same predictions, physicists accepted them as expressing the same physical theory.

By developing a mathematical formulation as powerful as Newton's description of the effects of gravity, Maxwell capped two centuries of exploration with an encompassing description of electricity and magnetism. The equations that make up Maxwell's model were based on a mechanical model of the nature of electromagnetism, which was abandoned after repeated failures to measure the velocity of the earth through the ether. Although the idea of a pond (the ether) was given up, the waves took on a life of their own. In the way physicists spoke, the electromagnetic field became an independent entity every bit as real as water.

A physicist constructs a model of a phenomenon—say the behavior of the roller coaster in our earlier example. Recall the physicist replacing the "flesh and blood" roller coaster with a set of numbers representing the speed of the cars and their height above the ground. This replacement is called a model, and its adequacy depends upon its use, which determines how accurate its descriptions must be. When the first model failed to explain the fact that the cars could not quite reach the same height each time, the physicist augmented the model by adding the effects of friction.

We can contrast this way of talking about models with the everyday way we talk about models of cars or airplanes. The amount of detail in these models (whether they have engines, whether these engines work, etc.) depends on how we want to use them, but essentially they are smaller- or larger-scale versions of a "real thing." It is tempting to think of scientific models in the same way, and indeed they are often thought of in this way. But the failure of attempts to measure the motion of the earth through the ether cast doubt on the possibility of constructing any sort of convincing mechanical model of the abstract relationships given by Maxwell's equations.

Like electromagnetism, gravity can be talked about in the language of a field. For example, we can say that the earth is surrounded by a gravitational field whose properties determine the way bodies behave in the immediate vicinity of the earth. The properties of this field are mathematically directly related to Newton's expression for the gravitational force exerted by the earth on the bodies.

The concept of a gravitational field allowed physicists to talk about gravity in a new way. They now described motion as the result of a particle's response to the gravitational field in its immediate vicinity, rather than the particle's reaction to a physically remote object. The concept provides an answer, of sorts, to the question, Where does the kinetic energy of the roller-coaster cars go when the cars slow down as they climb a hill? The answer is, into the field. Still, the field seems like such a mathematical abstraction that only another mathematical abstraction—energy—could possibly be stored there. Energy is an abstraction because it is not tangible (nor is the height of the roller-coaster cars). We can calculate the amount of energy, but we cannot isolate it or photograph it. Energy is an accountant's notion, like the national debt—real enough, but one cannot point to it or put it in a vault.

Physicists can define a gravitational field, but does a gravitational field have any reality beyond its definition? Is it something physical or is it a mathematical fiction? This distinction is not as clear-cut as it may at first appear.

The field description is apparently quite different from Newton's model of force, replacing the interaction of a particle with distant objects with an interaction with the field in the vicinity of the particle, but the underlying mathematics of the two approaches turns out to be identical. This identity guarantees that the field description will allow physicists to make exactly the same predictions Newton's approach allows. The gravitational field is just as real (or as unreal) as Newton's action at a distance. Remember that as far as a physicist is concerned, the important quality of a theory is the predictions it makes. Since both approaches make the same predictions, they are not two different theories.

In the words of Richard Feynman:

> Many different physical ideas can describe the same physical reality. Thus, classical electrodynamics can be described by a field view, or an action at a distance view, etc. Originally, Maxwell filled space with idler wheels, and Faraday with field lines, but somehow the Maxwell equations themselves are pristine and independent of the elaboration of words attempting a physical description. The only true physical description is . . . the way the equations are to be used in describing experimental observations.[14]

We feel we understand something when we can picture how the wheels and levers must fit together in order for it to work. Physicists were reluctant to abandon this level of understanding until they had no alternative. However, the development of physics in the twentieth century has been a progressive movement away from visualizable models and toward abstract mathematical models.

Physicists now talk of fields in a much more abstract way than Faraday or Maxwell did. A field is now thought of as a way of assigning numbers to a region of space, much as a temperature map assigns a temperature to every point on the earth's surface. Although this description makes a field seem very abstract, it proves to be a very rich way of talking about nature. In fact, physicists today talk about fields in exactly the same way as they talk about material objects. The story of twentieth-century physics is the story of how fields moved from Maxwell's mechanical description to an abstract mathematical description and finally, as we shall see in later chapters, to the "concrete." During the nineteenth century, atoms began the same journey.

4

The Ingenious Notion of Atoms

If, in some cataclysm, all of scientific knowledge were to be destroyed, and only one sentence passed on to the next generation of creatures, what statement would contain the most information with the fewest words? . . .[15]

RICHARD FEYNMAN

We humans have a deep-seated need to imagine an order behind the confusion that often seems to confront us. The gods were one of the earliest structures that people said were hidden behind direct experience. The gods' suspiciously human motives were said to explain occurrences as diverse as the weather and wars. But even people who did not talk about purposeful forces shaping the behavior of the world often called upon equally invisible mechanisms.

Over 2,000 years ago, the Greek philosophers Leucippus and Democritus championed the idea that the world is composed of particles of matter too small to be observed and that everything we find in the world arises from differing arrangements of these particles. Democritus held that nothing exists save these atoms and the void through which they move.

For most of human history there was as little evidence in support of the atomic view of matter as there was in support of the existence of the gods. Eventually the atomic view would be transformed from an unsupported philosophical speculation to a respectable scientific fact, but it would take more than two millennia for this process to unfold. Even as late as the end of the nineteenth century, some influential physicists opposed the atomic view of matter, but initial acceptance of the atomic theory had begun because of physicists' attempts to understand the behavior of gases.

The Behavior of Gases

Why can we compress and expand a gas when solids and liquids resist such changes? Physicists developed two ways of talking about the behavior of gases. In the first description, gases are made up of something that resembles tiny springs in contact with each other. These springs compress when the gas is compressed and expand when the gas expands. In the second description, the gas is made of particles ("atoms") that are far apart but move about rapidly, colliding with each other and the walls of the container in which the gas is held. When the gas is compressed the spaces between the atoms become smaller; when it expands the spaces between the atoms become larger. Physicists described the results of their experiments in both vocabularies, and both did reasonably well.

Newton's language made it possible to develop a more detailed description of gases, based on the idea that heat in a gas is the result of the random motion of particles of gas. In this description the pressure in a gas is produced by the collisions of the particles with the walls of the container. When the volume of a container is decreased, the number of impacts, and hence the pressure, increases.

Talking about the world in terms of indestructible atoms helped make sense out of many observations. For example, the French chemist Antoine Lavoisier argued in the eighteenth century that the same amount of matter is always present before and after each chemical reaction. Lavoisier based his conclusion on a series of experiments showing, for example, that when iron rusts it becomes heavier, but if the iron is placed in a closed container, the air becomes lighter by a compensating amount. The notion was to be elevated to the principle of the *conservation of mass*.

The Italian physicist Amedeo Avogadro, in the early nineteenth century, described gas in terms of atoms combined into molecules that are quite small in comparison to the distance between them. The molecules had to move rapidly to give the illusion of filling the space occupied by the gas. Avogadro conjectured that all gases at the same temperature and pressure have the same number of molecules in similar volumes. This is essentially the way that physicists and chemists talk about gases today, but Avogadro's approach was not widely accepted until the twentieth century.

The continued failure of the alchemists to transmute lead or iron into gold supported the notion that there are chemical elements with distinct properties and that these cannot be converted into one another. These elements could be characterized not only qualitatively but also quantitatively, in terms of the way different weights combine—for example, the constant ratio, by weight, of oxygen and hydrogen gases in combination to form water. The chemist John Dalton, around 1800, used such observations to argue forcefully for an atomic description of matter. Dalton's "combining weights" could be associated with the relative weights of the "atoms" involved, and were therefore called *atomic weights*.

In 1869 the Russian chemist Dmitri Mendeleev organized the current state of knowledge of the properties of the chemical elements into a form that demonstrated their similarities and differences based on each element's atomic weight and the "affinities" of different elements for combining with other elements. This periodic table of the elements, as it came to be known, was not completely filled, and the blanks pointed to the possibility of as yet undiscovered elements and provided chemists with clues about their properties. The promise of the periodic table was soon fulfilled as chemists discovered some of the new elements it had predicted. As valuable as the periodic table was, however, there was still no clue to what underlying mechanism allowed it to work.

The Nature of Heat

Newton's and Maxwell's laws describe in detail what is taking place, whether it is the motion of a planet, the behavior of an electromagnetic wave, or the action of atoms in a gas. Physicists found that they could talk about nature in another way, however, that did not require them to have any idea of the detailed processes that might be occurring. A good example of this way of talking is the expression of the conservation of energy outlined in the roller-coaster experiment.

The German physician Robert Mayer first proposed the conservation of energy in 1842 in order to explain his medical observations. He found that blood in the veins appears to be brighter in the tropics than in Europe. This led him to think about the methods animals use to generate heat and about the ways energy can be converted from one form to another. Mayer argued that although energy appears in many different forms—mechanical, electrical, chemical, and biological—and although these forms can be exchanged one for another, the total amount of energy in the universe remains constant. Mayer's formulation of the conservation of energy was very broad, philosophical, and qualitative, and physicists, who were by now used to the precision of Newtonian mechanics, paid it little attention.

For many years physicists talked about heat in two different ways. In the kinetic vocabulary, heat is produced by the motion of the particles making up a substance. This approach is obviously very similar to that taken in the kinetic description of gases. Other physicists identified heat with a substance, sometimes called *caloric*, that flows from one body to another. Like the kinetic description, the caloric description was compatible with the other things that physicists believed.

Some data, such as the observation that rubbing something causes it to get warm, seemed to support the kinetic, or motion, description of heat. Other observations—such as the fact that when bodies of different temperature are brought into contact with each other, heat seems to flow from the hotter to the colder body but is neither created nor destroyed in the process—seemed to favor the caloric description.

In the middle of the nineteenth century the Englishman James Joule carried out extensive measurements ostensibly designed to improve the efficiency of the engines used in his family brewery. On the bases of the results he obtained, Joule formulated a quantitative mathematical expression for the relationship between heat and mechanical energy and demonstrated the validity of this relationship in a wide variety of settings. Physicists were comforted to see the concept of heat dealt with in a quantitative way, and Joule's work advanced the acceptance of the idea that heat is a form of kinetic energy. The vocabulary of the caloric began to fall out of use.

The idea that heat is something that flows from a hotter to a cooler body is still very much a part of everyday language, and the language of physics still bears evidence of its origins. For example,

at the beginning of the nineteenth century the French mathematician Jean Baptiste Fourier developed equations to describe the flow of heat within the earth based on a view of heat that was similar to the caloric approach. Fourier's immensely powerful way of solving these equations has proved far more viable than the theory it was meant to support. Fourier analysis has become an almost ubiquitous mathematical tool in modern physics, but the kinetic theory proved a superior way to describe the behavior of heat.

Assume for the moment that everything is made up of atoms and that these atoms are in motion. If all the atoms are moving in unison the object is in motion. If the atoms are all moving in different directions, the object is warm. The faster the average random motion, the warmer the object. From this perspective the brakes on your car are simply devices to convert the energy contained in the motion of the car into energy in the form of heat. Something has to get hot (the brakes) if the energy associated with the motion of your car is to be conserved when you stop.

Meanwhile, quite a different approach to talking about heat was emerging. The French engineer Sadi Carnot, who was concerned with understanding the principles behind the operation of the steam engine, posed this question in 1824: What is the maximum efficiency with which heat can be converted into useful energy? His conclusion is embodied in what is now known as the second law of thermodynamics. (The first law would turn out to be the name given to our old friend the conservation of energy.)

The language of thermodynamics is even more fundamental than Newtonian physics. The laws of thermodynamics can be expressed in several different ways whose equivalence is not always apparent. For example, one way of expressing the second law is to say there can be no machine that gains its energy from a source colder than itself. The laws of thermodynamics can also be thought of as prohibiting the building of perpetual motion machines. The most general formulation of the second law is that the amount of disorder, or entropy, in the universe increases in time—in other words, other forms of energy are converted into heat.[16] Surely an abstract principle if there ever was one, this formulation gives us no clue about *why* other forms of energy are converted into heat or what detailed processes are involved.

Furthermore, the second law introduces an entirely new concept into physics.

The laws of mechanics as formulated by Galileo and Newton have no preferred direction in time. Run backward, a film of the motion of the planets shows them obeying Newton's laws just the way a film running forward does. Moreover, there is no way that we could tell whether a film of the motion of the planets *is* being run backward. On the other hand, we have no trouble recognizing a film running backward of an ice cube melting. It seems to show something we never see in the world—a puddle of water forming itself into an ice cube with no help from a tray or refrigerator. In other words, melting ice cubes, falling trees, and burning wood draw our attention to the direction of time. Time points from the more orderly (the ice cube) to the less orderly (the puddle). But how?

The first law of thermodynamics, the conservation of energy, can be expressed in Newtonian terms (physicists say the first law is more fundamental than Newton's laws, however, because it applies even in cases where Newton's laws do not). In view of the triumph of the Newtonian approach, physicists, reasonably enough, tried to express the second law of thermodynamics in Newtonian language. This proved to be much harder than many of them imagined.

In 1860, James Maxwell, of electromagnetic fame, developed a mathematical expression for the speeds molecules would travel at different temperatures. After considerable work attempting to create a Newtonian model of the second law, the Austrian physicist Ludwig Boltzmann, in the latter part of the nineteenth century, argued that the direction of time is not an absolute, but rather a *statistical* effect. Of all the possible outcomes, most have a greater degree of disorder than any particular arrangement we might be considering. If we wait long enough, we might see an ice cube form out of a puddle of water. It is not impossible; it is just very, very unlikely.

For many physicists such a statistical explanation was unsatisfactory. They were certain that the second law of thermodynamics must not be simply a statistical outcome but something more fundamental. Further progress on the question (it is still not completely resolved) had to await the invention of a more powerful description of the nature of matter on the atomic level.

Physics at the End of the Nineteenth Century

When physicists say they understand a phenomenon, they usually mean that they can write an expression describing quantitatively how the phenomenon unfolds in time. To the degree that understanding nature is demonstrated by the ability to describe and predict the behavior of physical systems, physics had made astounding progress by the end of the nineteenth century. Physicists could describe the behavior of bodies moving under the influence of gravity and mechanical forces with astounding precision. The effects of the forces of electricity and magnetism were equally predictable, and these forces were united in a single description predicting the existence of electromagnetic radiation and providing a powerful way to talk about the nature of light. Physicists knew the principles governing the operation of machinery and felt that these embodied fundamental principles that applied to the universe as a whole. And after more than 2,000 years, chemists seemed on the threshold of convincingly demonstrating the atomic nature of matter.

Beyond the realm of description, the situation was a lot less clear. Two hundred years after the publication of the *Principia,* physicists were able to use Newton's formula to predict the effects of gravity, but they were no closer to understanding the nature of gravity than Newton had been. Furthermore, there were apparently very different ways of describing the behavior of systems. These descriptions had the same underlying mathematical structure and hence made the same predictions; we could ask which of them more closely described nature, but there seemed to be no way to tell—nor did the answer seem to make any difference.

Even the idea of what it means to be a description of nature was shifting. The ground had been cut from under the mechanical model that attempted to explain electromagnetic waves, and physicists had no idea what these waves might be or how they could possibly propagate through space. The most powerful generalization physicists knew was easily formulated: "The energy of the universe remains constant, while its entropy increases." But this formulation did not give them the slightest hint of how or why energy stays constant or entropy increases.

Yet who could argue with the immense success of physics? *Why* was a question that could be left for philosophers to grapple with; physicists were happy to settle for increasingly precise descriptions and the power of prediction. There were still some unanswered questions, but no one could foresee the revolution in outlook necessary before these questions could be answered. The turn of the century brought the beginnings of this revolution.

What about the quotation with which this chapter began, about the one sentence that contains the most scientific information with the fewest words? Feynman suggests:

> *All things are made of atoms—little particles that move around in perpetual motion, attracting each other when they are a little distance apart, but repelling upon being squeezed into each other.* In that one sentence . . . there is an enormous amount of information about the world, if just a little imagination and thinking are applied.

5

The Unimaginable Unity of Spacetime

No one must think that Newton's great creation can be overthrown in any real sense by this or any other theory. His clear and wide ideas will forever retain their significance as the foundation on which our modern conceptions of physics have been built. [17]

ALBERT EINSTEIN

After what seemed like an interminable wait, a recent graduate of the Swiss Technological Institute landed his first job as a patent examiner in Bern, Switzerland. In addition to his new duties, Albert Einstein managed to write and publish three papers in 1905 that shook physics to its foundations. One provided a demonstration of the atomic nature of matter, and the other two revolutionized the way physicists talk about space, time, and light. We will look briefly at two of these papers in this chapter and the third in the next chapter.

In the first paper, Einstein argued that the existence of atoms should be betrayed by the motion of extremely small particles suspended in a gas. Although atoms are too small to be observed directly, Einstein said that the effect of the collisions of atoms with the small particles could be observed in the random motions of the particles, and he used the kinetic model of Boltzmann to predict the extent of this observable motion. The effect Einstein described was already familiar to biologists who used microscopes to study tiny organisms, although the biologists did not understand what they were seeing. The random motions of tiny particles in a liquid were named after the man who first described them in 1827, the English botanist Robert Brown. At first Brownian movement was thought to be biological in nature, but Einstein's quantitative description provided, for many, a final piece of evidence in support of the atomic theory. This paper also demonstrated a quality that would become increasingly important in physics—indirect evidence. Einstein produced evidence for the existence of atoms by calculating an effect produced by these invisible entities.

Special Relativity

Einstein was deeply influenced by the success of thermodynamics. He was attracted to the idea of a theory based on a few powerful principles that do not have to rely on an underlying mechanical model, and he took this approach in his paper on what has come to be called special relativity. He wrote this paper, not to explain a new phenomenon, but to describe what seemed to him to be a deep simplicity at the heart of nature.

Michael Faraday, the inventor of the idea of the field, along with the American physicist Joseph Henry, discovered that a moving magnet generates an electric current in a coil of wire surrounding the magnet. If instead of moving the magnet, the coil of wire surrounding the magnet is moved, a current is also produced in the coil. The two processes were distinct in the interpretation of Maxwell's equations, but Einstein was sure this distinction must be artificial—he believed that only the relative motion between the magnet and the coil should be important. To eliminate the artificial distinction he saw describing in Maxwell's equations, Einstein put forth two principles, on which he then developed an entirely new way of talking about space, time, matter, and energy.

Einstein's first principle is that the laws of nature must look the same to any two observers whose speed and direction are not changing. In other words, no experiment performed by such observers is capable of telling which of two objects, such as the magnet and the coil of wire, is "really" moving. This principle is a generalization of the relativity of motion first articulated by Galileo. Newton retained Galileo's way of talking about motion, maintaining, as Galileo doubtless would have, that motion must take place against a motionless background—motion, in other words, is absolute. Einstein, however, argued that we never observe such a motionless background, and in any case, it seems to have no effect on what we do observe. For these reasons, Einstein said it is pointless to talk about a standard of absolute rest. He then went on to apply the vocabulary of the relativity of motion to electromagnetic as well as to mechanical systems.

Galileo's claim that the earth moves is rendered moot by Einstein's first principle. From Einstein's perspective, Galileo did not take his own arguments on the relativity of motion seriously enough. The earth's motion is no more fundamental than the sun's; it depends upon one's point of view. The notion that the earth moves around the sun leads to a much simpler view of a nature harmoniously reflecting Newton's or Einstein's laws, but as for which "really" moves, Einstein said there could be no answer.[18]

Einstein's second principle is equivalent to the statement that the speed of light in empty space will appear to be the same for all observers moving at constant speed and not changing direction, no matter how fast they are traveling with respect to each other. From Newton's perspective, such observers are experiencing no external forces and are therefore moving in accordance with his first law of motion; such observers can be called *inertial,* since Newton's first law is also called the law of inertia.

Einstein elevates the fact that the speed of light is always measured to have the same value no matter how fast its source is moving to a basic assumption or fundamental principle. This principle in turn implies that Maxwell's equations apply for all observers. The reason is that in Maxwell's system the speed of light can be calculated using measurements of the strengths of the electric and magnetic forces. If the relationship between the strengths of the two forces differed for inertial observers moving with respect to each other, the laws of physics could not be the same for these observers and the observers could use this difference to tell which of them was "really" in motion. By maintaining that the velocity of light in empty space must be the same for all inertial observers, Einstein was postulating that the laws governing electromagnetism must also be the same for these observers.

The constancy of the velocity of light flies in the face of our experience of everyday things. The cup that falls from our tray in an airplane seems to us to have little forward motion, but to an observer on the ground it seems to be moving at 500 miles an hour along with the airplane. Einstein tells us that if instead of dropping a cup, we turn on a flashlight, the speed we measure for the light as it travels to the front of the plane would be exactly the same as the speed measured by an observer on the ground—it would not depend on the speed of the source or of the observer. In this respect light seems to resemble nothing in our experience—particle or wave. Yet the experimental evidence supports Einstein's position: No matter what the relative speed of the source and the observer, light always is measured as having the same speed.

Einstein's second principle gives us another reason that Michelson and Morley could not measure the earth's motion through the ether—the velocity of light does not depend on the motion of the source or the target. Rather than explaining the failure to detect the motion of the earth through the ether, this principle elevates that failure to the status of a fundamental assumption. The ether is superfluous because it has no observable consequences. It might seem that the theory of relativity does away with the need for an ether by making it unobservable; however, unobservable entities will come to play a surprisingly fundamental role in physics. Unobservable *consequences,* however, eventually doomed the ether.

Saying the speed of light in a vacuum is the same for all inertial observers, and at the same time saying the laws of physics must be the same for these observers, leads to unexpected and surprising consequences. First, it requires giving up the firm conviction that space and time are two distinct entities. The speed of light cannot be the same for all observers unless space and time are *not* completely different kinds of things but rather can be traded one for the other. Clearly, space does not even resemble time, so how can we exchange one for the other? Relativistic effects become apparent only at velocities close to the speed of light, speeds we normally do not experience. Still, the idea that space and time are fundamentally the same violates our intuitions. The idea violated many physicists' intuitions as well, and Einstein's arguments were slow to be accepted. The ether, which had proved so useful in leading to the acceptance of the language of fields, became an obstacle to the acceptance of the language of relativity. When Einstein won the Nobel Prize for 1921, his development of the theory of relativity was pointedly not cited.

If we are rapidly moving away from each other, your clock will appear to me to be running slower than mine and mine will appear to be running slower to you. If we ask whose clock is *really* running slower, we find the question cannot be answered. To answer the question, we would have to be able to tell which of us was *really* moving, and the first principle of relativity—that the laws of nature are the same for all inertial observers—tells us we can never perform an experiment to determine this. Rather, the apparent rate at which a clock runs must always be related to the point of view from which the clock is observed.

The slowing down of clocks is well known to physicists who work with subatomic particles that move at close to the speed of light. Some of these particles break up or decay in very short intervals of time. However, when the particles are moving almost at the speed of light with respect to the laboratory, they can appear to live 10 or 20 times longer than when they are moving much more slowly.[19]

Einstein wanted to abolish the absolute distinction between space and time. To be mathematically consistent, he had to abolish the absolute distinction between matter and energy—to invent a way so that matter can be described in the vocabulary of energy and vice versa. This interchangeability seems to play havoc with our conviction that matter is some kind of "stuff," unless energy is some kind of "stuff" too. Unfortunately, as we have seen, energy is more an abstract notion than any kind of stuff. To talk about matter and energy in the same way, matter must be equally abstract.

In the fifth century B.C., the Greek philosopher Heraclitus proclaimed that everything was made of fire. As Werner Heisenberg, one of the founders of the twentieth-century revolution in physics, said, "Modern physics is in some way extremely near to the doctrines of Heraclitus. If we replace the word "fire" by the word "energy" we can almost repeat his statements word for word from our modern point of view."[20]

Today there is no question about the fundamental importance of Einstein's way of talking about space and time. The atomic and hydrogen bombs provide convincing evidence of the power of talking about the interchangeability of mass and energy. In Einstein's language, from our point of view, objects seem to gain mass as their velocity approaches the speed of light. In Newtonian physics, no such thing happens. If Newton's language described nature better than Einstein's, particle accelerators would look very different than they do. For example, instead of the two miles it takes to accelerate electrons to their highest energy in the Stanford Linear Accelerator, it would take only one inch!

Even before the idea of the conservation of energy was developed, Lavoisier argued, on the basis of his study of the rusting of iron, that matter can be neither created nor destroyed. The success of the special relativity theory predicts, although this way of talking fits everyday experience, the conservation of mass does not apply in every case. Matter is not always conserved—matter can be created and destroyed. Nevertheless, once again the remarkable adaptability of the way physicists talk about the conservation of energy comes to the rescue. If, whenever physicists talk about energy, they include both the traditional forms of energy and the energy locked up in the form of matter according to Einstein's famous equation $E = mc^2$, then they can again say that energy is conserved. In Einstein's hands, the notion of the conservation of energy underwent a metamorphosis and emerged even more powerful than before.

Since matter can be described as energy, and since energy does not seem to be some kind of stuff, it seems that the universe is not made up of any kind of stuff at all. Then what *is* it made of? This is the province of atomic physics, and in this area the second great revolution of twentieth-century physics occurred. But before we examine this development, and the role played in it by Einstein's third historic paper of 1905, we will look briefly at the general theory of relativity, developed by Einstein in 1916, in which he invented an entirely new way of talking about gravity.

General Relativity

The general theory of relativity extends the vocabulary of relativity to a discussion of the behavior of gravity. In developing his theory, Einstein once again made use of something that had been obvious to physicists for some time but whose significance only he had the insight to see.

Galileo maintained that all objects fall to earth with the same acceleration. Newton declared the acceleration given to an object depends upon its mass and the force exerted on it. Therefore the force exerted by gravity on any object must be related to the object's mass. While electricity and magnetic forces depend upon the makeup of the objects involved, gravity does not. Why? How does gravity know what the mass of an object is, and why should gravity care about an object's mass, if the final result is simply that all objects fall with the same acceleration? Einstein answered these questions by a radical departure from Newton. Einstein gave up the approach of talking about gravity as a force.

We don't feel a gravitational force acting on us as we go about our daily activities. Rather than being "pulled" to the earth when we fall, it seems more like we are following a path of least resistance. The "force" of gravity is a convenient way of describing why objects do not move in straight lines when they are in the vicinity of massive objects like the earth, but perhaps there are other ways of talking about this behavior.

On the surface of the earth, the shortest distance between two points, as the crow, or rather as the airplane, flies, is some segment of a great circle. A great circle is a path on the surface of the earth traced by a slice through the center of the Earth. The lines of longitude are great circles. Imagine two travelers who set out, at the same time, traveling north from different points on the equator. They are on parallel paths, but as they approach the North Pole, they will come closer and closer to each other. We could even say they are being attracted to each other by a force that grows stronger the closer they get. But of course there is no need to talk about such a force. Each traveler is simply moving along the shortest route to his or her objective.

We can talk about gravity in exactly this way, Einstein said. We can maintain that all objects follow the shortest path from one point to another. In the absence of what we call a gravitational field, the path is the straight line given by Newton's first law. In the presence of a gravitational field, objects still travel along the shortest path, but the path is now curved. This curvature is analogous to the curvature of the surface of the earth, but according to Einstein, in the presence of a gravitational field, the geometry of space and time itself is warped by matter. The curved paths of the planets require no special mechanism—the planets are traveling along the shortest paths.

If we recall the example of throwing the ball upward, we realize that Einstein has no need to talk of forces. The ball always takes the shortest path, whether in space far from any massive body or in the vicinity of a planet or star. Whether or not we use the language of force is not important. What is important is how the language we use allows us to predict the ball's path. In this sense, Newton's and Einstein's languages serve equally well in most situations.

Just as the special theory of relativity is based on the assumption that it is impossible to perform an experiment to measure uniform motion, the general theory is based on the assumption that it is impossible to perform an experiment to tell whether we are falling freely in a gravity field or floating in space far from any massive body.[21] We are familiar with this phenomenon as we watch the astronauts float in "zero G" while they are falling around the earth at 18,000 miles per hour.

In the process of developing the general theory, Einstein had to make use of a new vocabulary—a new variety of mathematics. Unlike Newton, however, who was forced to develop the language of differential calculus, Einstein discovered that the mathematics of tensor analysis and differential geometry he needed to formulate general relativity had already been developed by the nineteenth-century German mathematicians Karl Gauss and Georg Riemann and by the turn-of-the-century Italians Gregorio Ricci-Curbastro and Tullio Levi-Civita.

The predictions of Einstein's general theory of relativity differ only subtly from those of Newtonian gravity on the scale of everyday experience, but there is one exception in the solar system. For some time astronomers had worried about a discrepancy between the Newtonian prediction and the observed motion of the planet Mercury. Astronomers assumed that there had to be an undiscovered planet closer to the sun than Mercury whose presence affects the orbit of Mercury. Despite intensive search for the planet, which they called "Vulcan," it was never found.

The discrepancy is a difference between the observed and predicted locations of the point in its orbit at which Mercury is closest to the sun. At the end of a century, the observed position differs from the position calculated using Newton's theory by 43 seconds of arc, roughly 2 percent of the angular diameter of the moon. Not a very large amount by most standards, but Einstein called for just this difference from the Newtonian prediction. This marvelous agreement convinced Einstein that he had found a powerful new way to talk about gravity.

The evidence that convinced most of the scientific world of the value of Einstein's approach came in the form of photographs of the background of stars visible near the sun at the time of a total eclipse. Einstein predicted that the light of distant stars should be deflected by a massive body such as the sun. In 1919 the English astronomer Sir Arthur Stanley Eddington compared photographs taken during an eclipse with photographs of the same region of the sky taken many months earlier, when the sun was in another part of the sky. The difference in the apparent locations of the stars when the sun was present and when it was absent allowed Eddington to determine how much the mass of the sun deflects the light of distant stars. The observed deflection agreed with Einstein's prediction.

Our first thought might be that just as the kinetic description of heat replaced the caloric description, so general relativity replaced Newton's language. But this conclusion is much too hasty. To this day the overwhelming majority of calculations involving the effects of gravity use Newton's model and not Einstein's. Why? Because Newton's approach is completely adequate even for such complex tasks as guiding distant spacecraft through Saturn's rings. Only in special cases, such as making extremely precise measurements in the solar system, calculating the effects of such conjectured massive objects as black holes, and trying to understand the universe as a whole, is Einstein's theory needed.

Although physicists view Einstein's approach as more fundamental than Newton's, Einstein's vocabulary did not replace Newton's. In Heisenberg's words, "Wherever the concepts of Newtonian mechanics can be used to describe events in nature, the laws formulated by Newton are strictly correct and cannot be improved."[22] A drill press may do many things a hand drill cannot, but as far as a carpenter building a house is concerned, the drill press is not necessarily a better tool. Likewise, whether physicists think of gravity as a Newtonian force or as an Einsteinian distortion in the fabric of space and time depends on the problem they are dealing with. Moreover, we shall see that there are still other ways of talking about gravity.

To a degree, Einstein's general theory of relativity restores the ether banished by the special theory. Spacetime in the general theory has a variety of properties, such as stress and energy density, that seem to require the existence of something not very different from the ether. Quantum mechanics also attributes a variety of properties to what would otherwise be empty space. Nevertheless, the mechanical, visualizable model of the ether is gone, apparently never to return. It was replaced by abstract ether, well suited to mathematical expression but far from what Maxwell had in mind when he first described electromagnetic waves.

Galileo's accomplishment was made possible by his decision to talk about the world in terms of motion through space and time. These concepts seem so obvious to us that it is difficult to remember that they *are* concepts. Time is normally measured in terms of motion, from the swing of the pendulum of a grandfather's clock to the oscillations of a quartz crystal in a modern watch. Apart from such periodic behavior, how could we even talk about the uniformity of time? In the words of the contemporary American physicist John Wheeler, "Time is defined so that motion looks simple." Wheeler also said, "Time? The concept did not descend from heaven, but from the mouth of man, an early thinker, his name long lost."[23]

Einstein demonstrated the power of talking about space and time as though they were a unity, and in the process he showed that both space and time are human inventions—ways of talking about the world.

The Imponderable
Nature of Matter

A new scientific truth does not triumph by convincing its opponents and making them see light, but rather because its opponents eventually die, and a new generation grows up that is familiar with it.[24]

MAX PLANCK

In 1897 the English physicist Joseph J. Thomson was studying the behavior of what would prove to be the ancestor of the television tube. A cathode-ray tube, as it is called, consists of two metal plates mounted at opposite ends of a glass tube from which most of the air or other gas has been removed. If each plate is connected to one of the two terminals of a high voltage electric source, a glow will appear at the end of the tube near the plate connected to the positive terminal.

Thomson reasoned that something must be leaving the plate connected to the negative terminal and striking the tube near the positive terminal, producing light. He found this something could be deflected by both electric and magnetic fields, leading him to conclude that whatever the something is, it must carry electric charge. Since he got the same results no matter what material he used for the negative terminal, he felt whatever was producing the glow must be a universal constituent of matter.

By studying the differences between the effects of magnetic and electric fields in deflecting the particles, Thomson was able to compute the speed of the particles, which proved to be astoundingly high—one-tenth the speed of light. Furthermore, he could calculate the mass associated with each unit of charge. Once again the answer was a surprise—the mass associated with each charge was 2,000 times smaller than the same relationship for the nucleus of an atom of the lightest element, hydrogen. Thomson had discovered a new constituent of the world, one much less massive than the atom—a particle he named the electron.

A New Vocabulary

In December 1900 the German physicist Max Planck ushered in the twentieth century by reading a paper that resolved a long-standing problem for physicists. When we heat an object in a stove until it glows, it gives off most of its energy at frequencies characteristic of red light, which is why we see a red glow. If we continue to heat the object, it will glow orange, yellow, and finally white hot. The theory of heat developed by Maxwell could not account for this well-observed behavior. This failure was a challenge to physicists. No one imagined the price that would have to be paid to solve this puzzle.

Physicists had learned to identify heat with the energy associated with the motion of molecules. Analogously, Planck reasoned that the radiation emitted by a body could be said to be produced by the vibration of tiny oscillators, which might or might not be atoms; these oscillators move because of the internal energy associated with the temperature of the body. Planck was able to invent a formula correctly describing the relationship between frequency and intensity for the radiation given off by a heated body. When he tried to derive this formula from more basic and previously accepted concepts, however, he failed. He was able to succeed only by making the arbitrary assumption that the oscillators could exchange energy with the electromagnetic field only in discrete bundles. The size of these bundles is determined by the frequency of the radiation and by a new constant of nature that Planck called the *elementary quantum of action*. Until this time there was no reason to assume the existence of discrete units of energy, but try as he might, Planck could not eliminate this strange requirement. Planck described the situation in his acceptance of the Nobel Prize in 1918:

> Either the quantum of action was a fictional quantity, then the whole deduction of the radiation law was in the main illusionary and represented nothing more than an empty nonsignificant play on formulae, or the derivation of the radiation law was based on a sound physical conception. In this case the quantum of action must play a fundamental role in physics, and was something entirely new, never before heard of, which seemed called upon to basically revise all our physical thinking,

built, as this was, since the establishment of the infinitesimal calculus by Leibniz and Newton, upon the acceptance of the continuity of all causative connections.[25]

Despite the wrenching consequences of accepting Planck's analysis, the agreement between predictions based on Planck's formula and the intensities actually measured at different wavelengths was too good to ignore.

Einstein linked Thomson's electron with Planck's model. Einstein had the vision to see that the relationship between energy and frequency demanded by Planck's formula could be used in talking, not only about oscillators in matter, but also about light itself. He used this approach in his analysis of a phenomenon known as the photoelectric effect in his third epoch-making paper of 1905. When light falls on certain metals, Thomson's electrons are emitted from the surface of the metal. The German physicist Philipp Lenard found that no matter how intense the light is, its frequency must be higher than a particular frequency characteristic of each metal or no electrons will be released. Red light, which is characterized by oscillations at relatively low frequencies, produces no electrons; blue light, with its relatively higher frequency, causes electrons to be released by the metal. This difference poses a problem for models based on Maxwell's theory, because a sufficiently intense red light should agitate the electrons more than a weak blue light does, and thus the red light ought to shake electrons free. Why then does red light produce no electrons?

Einstein applied Planck's argument that the amount of energy in an interaction involving radiation is given by a number, represented by the letter h (Planck's constant), times the frequency of the radiation involved. Since the frequency of blue light is higher than the frequency of red light, there is more energy in each interaction of the metal with blue light than there is in each interaction with red light, and hence a greater opportunity to knock an electron free from the surface of the metal. Einstein argued that the photoelectric effect demonstrates that radiation interacts with metal as though *all* of the energy in radiation is concentrated in "packets" of energy, which we now call *photons*. Radiation in this experimental arrangement behaves as though Newton had a better way of talking about light (as particles) than Huygens and Maxwell did (as waves).

In Einstein's description, an electron interacts with only one photon at a time. Even though there are many packets of energy in a bright red light, each packet lacks sufficient energy to knock an electron free. While a dim blue light has fewer packets of energy, each packet has enough energy to free an electron. The frequency of the light determines whether *any* electrons are freed. If electrons are emitted, then the intensity of the light determines how many electrons are knocked loose. Einstein predicted a relationship between the current produced by the electrons emitted from the metal and the intensity of the light falling on the surface. At first experiments seemed to be at variance with Einstein's predictions, which were not experimentally supported until 11 years later, because the measurements were difficult to make.

Einstein's paper was greeted with less than unbridled enthusiasm. In fact, most physicists reacted to it with outright hostility. First, not much was known about the photoelectric effect and physicists thought it was much less fundamental than the radiation problem tackled by Planck. Second, Einstein seemed to be reinstating the discarded Newtonian description of light. He had done nothing to address the glaring conflict between his proposal that light is a particle and the overwhelming evidence that light is a wave. How could these two pictures be reconciled? Consistency is vital if any system is to make sense; Einstein was talking in a way that was apparently at odds with Maxwell's remarkably successful way of describing the behavior of electromagnetism. Even Planck, the father of the notion of the quantum of action, rejected Einstein's interpretation because it seemed incompatible with the overwhelming success of Maxwell's equations; Planck was convinced Maxwell had correctly described the behavior of light as a wave in space.

Einstein himself was deeply troubled by the conflict between the idea of light quanta and the evidence for the wavelike nature of light. He emphasized the provisional nature of his analysis, but he never stopped believing that Planck's vocabulary was essential for describing the behavior of light.

You may recall that Young's two-slit interference experiment (page

41) helped convince physicists that light must be a wave phenomenon, because the *same* wave front can pass through both slits and produce the interference. Now Einstein was saying that light is a particle. But clearly a single particle *cannot* pass through both slits at the same time. How then can a particle model explain the diffraction pattern? This apparently irreconcilable conflict lay at the heart of the refusal of most of the physics community to accept the idea of light quanta even *after* Einstein's formulation of the photoelectric law was accepted.

Einstein was awarded the Nobel Prize in 1922 "for his services to theoretical physics and especially for his discovery of the law of the photoelectric effect." But although Einstein had developed a formula describing the photoelectric effect, the citation did not even mention the quantum model upon which the formula was based—it was still too radical. Einstein's formula was accepted, but his explanation of what the formula entails—that light has a particlelike nature—was not acknowledged.

Evidence continued to mount, however, and ultimately physicists accepted the particlelike quantum nature of light. The decisive step probably came in 1922 with the American physicist Arthur Compton's demonstration that when X-rays (X-rays are electromagnetic radiation like light, but they have much shorter wavelengths) strike targets, they behave just the way they would if they were individual packets of energy. Physicists had become so desperate about reconciling the wave-particle conflict that some of them even considered abandoning the conservation of energy when talking about individual subatomic interactions, but Compton's work confirmed that the conservation of energy was still a sensible way to talk about events on the atomic scale.

Nevertheless, the paradox remained. Physicists now had two incompatible ways of talking about light: a wave vocabulary that applied in empty space, and a particle vocabulary that applied whenever light interacts with matter. They patched the two languages together, but the inconsistent expressions always left them uncertain whether their calculations would make any sense or not. Much of the effort of physicists in the 1920s was devoted to finding a way to combine the predictive power of both theories and to provide a unified way of talking about the behavior of light.

Talking about Atoms

Meanwhile, experimenters were exploring ways to study the still-invisible atom. At the beginning of the nineteenth century the Bavarian scientist Joseph von Fraunhofer studied the light emitted by heated gases. By passing the light through a prism and separating it into its constituent frequencies, Fraunhofer showed that heated gases give off radiation at definite frequencies. These frequencies turn out to be characteristic of the gas and of the temperature to which it is heated.

The kinetic theory of gases is based on a description in which atoms are perfectly elastic and similar to ideal billiard balls in many respects. As well as this model works in describing many of the physical properties of gases, it provides no clue to why the light given off by incandescent gases displays the particular frequencies observed by Fraunhofer. Clearly a model of the atom was needed that is more complex than a structureless, perfectly elastic, billiard ball.

The nature of the atom became even more enigmatic with an accidental discovery in 1896 by the French scientist Henri Becquerel. Becquerel had earlier noticed that the element uranium is capable of fogging a photographic plate. At first he thought some form of phosphorescence was involved. To test this idea, he placed a key between a piece of rock containing uranium and a photographic plate wrapped in black paper and exposed the package to sunlight to stimulate the phosphorescence. When he developed the plate, sure enough, he found a shadow image of the key. Since sunlight alone will not form such an image when the plate is securely wrapped, Becquerel attributed the effect to a form of phosphorescence in which the energy of sunlight is absorbed by the rock and then emitted, forming the image of the key. He was amazed to discover that the effect was even stronger when the package had spent several days in a drawer where there was no sunlight. The image was apparently produced by something given off by the rock itself. Furthermore, Becquerel found that the intense emission diminished only very slowly with time.

Other "radioactive" elements were discovered, and in 1903 the

French chemist Pierre Curie found that one gram of radium could melt slightly more than one gram of ice and convert it to steam in only one hour. Where was all this energy coming from? Does radioactivity violate the conservation of energy? What, if anything, does it say about the nature of the atom?

Ernest Rutherford, a physicist born in New Zealand who emigrated to Canada and then to England, carried out a series of experiments at the beginning of the twentieth century that convinced him that radioactivity realizes the dream of the alchemists: It transforms one element into another. (Unfortunately, however, rarer elements such as radium are transformed into more common elements such as lead—not exactly what the alchemists had in mind.) Rutherford's conclusions were not warmly received. The atomic model had only recently gained wide acceptance, and here was Rutherford telling physicists and chemists that the atom was not an immutable fundamental constituent of matter after all.

Rutherford found that radioactive atoms give off two types of particles, which he named *alpha* and *beta* particles. In a remarkably elegant experiment typical of his approach to answering physical questions, he demonstrated that alpha particles are related to helium atoms. He placed radium in a glass tube with walls that were not thick enough to stop alpha particles, then placed this tube inside a larger tube with thicker walls. After several days Rutherford passed an electric current through the larger tube and observed light of the frequencies characteristic of helium. The radioactive decay of radium produced helium. Questionable though his model of the transformation of the elements might have been to many physicists, Rutherford's experimental results were unassailable.

But how is the electron discovered by Thomson related to the atom? Thomson said electrons are probably embedded in a larger "pudding" of positive electricity. Rutherford designed an experiment to probe the structure of the atom and verify Thomson's model. Using the alpha particles whose nature he had uncovered, Rutherford studied the atomic structure of a thin gold foil. With the positive charge of the atom spread out thinly in space, he expected the alpha particles to encounter no more resistance than tissue paper would offer a bullet. Rutherford found most of the alpha particles did just what he expected them to do. A few, however, seemed to bounce backward upon encountering the foil—something Thomson's model could not explain. Rutherford realized that this observation provided an important clue to an improved model of the atom that was unlike Thomson's—one in which the atom is mostly empty space. The empty part is the part that the alpha particles pass through. The part of the atom that is not empty space, however, is massive enough that alpha particles bounce off it the way a rubber ball bounces off a brick wall.

Rutherford's observations led to the replacement of Thomson's model because they seemed to better fit a model of the atom resembling the planetary system, in which the nucleus of the atom plays the role of the sun and the electrons play the planets. This model is still the way most of us talk about atoms. In Rutherford's experiment the very light electrons are unable to deflect the much heavier alpha particles, which are four times as massive as hydrogen atoms. However, the nucleus of the gold atoms in the foil is much more dense. In addition, the nucleus carries a large positive charge. Because alpha particles are positive, they are repelled when they get close to the nucleus of the atoms and appear to bounce off the positively charged core of the atom.

The picture of a dense positively charged core of the atom surrounded by negatively charged electrons poses a real problem for physicists—it is unstable. As a negative electron circles a positive nucleus, the electron is constantly changing direction. Maxwell's equations call for an electric charge that behaves in this way to radiate energy. In fact, an electron should radiate away all its energy in a tiny fraction of a second and spiral into the nucleus. Atoms should collapse, and matter, as we know it, should not exist. With all the success of classical physics, it was easy to lose sight of the fact that the languages of Newton and Maxwell provide no way to describe the stability of the world. The most obvious property of the world—its persistence in time—was still not understood by physicists.

* * *

A New Language Is Born

The Danish physicist Niels Bohr proposed to solve the problem of atomic stability by simply ignoring the consequences of Maxwell's model. In 1913, Bohr declared, without offering any explanation, that contrary to Maxwell's description, when an electron is in orbit around an atomic nucleus it does *not* radiate energy at all. Bohr postulated that the only time an electron radiates energy is when it "jumps" in an atom from one orbit to another orbit. Each orbit is associated with a particular energy; when an electron jumps from one orbit to another, it emits a photon whose energy is the difference between the two orbital energies. The amount of energy in the photon emitted by the electron is related to Planck's constant, as is the energy of the photon in Einstein's model of the photoelectric effect.

Bohr had finally found a way to talk about the atomic nature of matter. He declared that only certain atomic orbits are permitted, just as Planck had made the equally unsupported declaration that oscillators in matter emit energy in discrete packages. Bohr also asserted that there is a lowest permitted orbital energy, so the electron cannot fall into the nucleus, as it could in a model based on Maxwell's equations. Furthermore, he said that *when* any particular electron will jump from one orbit to another is not determined—it is a matter of probability. Subatomic probability, ironically enough, was introduced by Einstein, who was later to become one of the last great physicists to oppose the idea that probability is a fundamental aspect of the physical world.

Bohr's way of talking about the atom also provided a way to explain why atoms in a gas do not lose energy when they collide—why they behave like perfect billiard balls. When one atom strikes another, unless there is sufficient energy to raise an electron in one atom or the other to a higher energy orbit, no energy at all is exchanged. This means that the atoms do not "heat up" internally, the way actual billiard balls do ever so slightly when they are struck. Despite his radical assumptions, Bohr's way of talking about atoms provided a way to calculate the frequencies of the radiation emitted by the hydrogen atom—and these calculations agreed with the observed frequencies.

Planck, Einstein, and Bohr each confronted a problem that would not yield to conventional analysis. Each of them proceeded to talk about the world in a way that was unsupported by the existing language of physics. The step each suggested was eventually accepted by other physicists, but only because it allowed them to solve problems with which the older approach could not cope. Yet there still remained the challenge of bringing these components into one consistent language. Bohr and others were aware that his model of the atom was only a step toward the desired solution. It still left too many answers beyond reach.

Doctoral students may dream that their thesis work will win them a Nobel Prize, but as far as I know only one person has achieved that dream—the French physicist Louis de Broglie. In 1924 de Broglie took the language of Einstein's paper on the photoelectric effect and carried it a step further. He argued that if light can be talked about in terms of a particlelike as well as a wavelike nature, why not talk about matter as having a wavelike as well as a particlelike nature? Using an analogy to Einstein's expression for the energy of a photon, de Broglie asserted that an electron in an atom has an associated wavelength, equal to Planck's constant, h, divided by the electron's momentum. Once this assertion was made, de Broglie readily showed that Bohr's atomic orbits are exactly those orbits that permit a whole number of electron wavelengths to fit the circumference of each orbit.

If a whole number of wavelengths exactly fit the circumference of an orbit, the waves overlap each other in such a way that crests fall on crests and troughs fall on troughs—the waves are reinforced. But if an orbit could contain fractional wavelengths such as 3.5, the waves would overlap in such a way that there would be crests and troughs at the same point in the orbit—the waves would cancel themselves out. Bohr's orbits no longer seem quite so arbitrary if electrons have a wavelike nature. But is there any evidence that electrons have such a nature? In other words, is de Broglie's assertion consistent with observations?

In 1927 two American physicists, Clinton Davisson and Lester Germer, carried out a series of studies in which electrons reflected from the surface of a nickel crystal. The electrons produced patterns that clearly resembled the diffraction patterns displayed by light in

Young's two-slit experiment more than a hundred years before. How could a particle behave as though it were spread out in space like a wave? Davisson described the problem in the following way:

> We think we understand the regular reflection of light and X-rays—and we should understand the reflection of electrons as well if electrons were only waves instead of particles. It is rather as if one were to see a rabbit climbing a tree, and were to say, "Well that is rather a strange thing for a rabbit to be doing, but after all there is really nothing to get excited about. Cats climb trees—so that if the rabbit were only a cat, we would understand its behavior perfectly."[26]

In 1928 the Japanese scientist Seishi Kikuchi photographed the diffraction of a beam of electrons passing through a thin sheet of mica. The wavelength of these electrons could be calculated from de Broglie's formula. Kikuchi then performed the experiment again using X-rays of the same wavelengths as the electrons. The pattern produced by the electrons and the X-rays was identical. De Broglie was right: Matter and light could be described using the *same* wave vocabulary.

While Bohr and others were working to build models of atoms, in 1925 Werner Heisenberg took a radically different approach to talking about atomic structure. The Austrian physicist Ernst Mach argued that only observable quantities were important. Heisenberg said that the important aspects of the atom are those that can be observed rather than any unobservable structure of the form Bohr sought to describe. Without attempting to describe the structure of atomic orbits, Heisenberg developed a mathematical approach to calculating the frequencies of the radiation observed to be emitted by an atom. Mathematicians were familiar with the mathematical tool Heisenberg developed—it embodied the rules for manipulating tables of numbers called matrices, and Heisenberg's technique came to be called *matrix mechanics*.

At the same time that Heisenberg was developing his way of calculating atomic properties, the Austrian physicist Erwin Schroedinger was working in a direction that seemed to preserve the traditional approach of Maxwell. In 1926 Schroedinger published a paper in which he described the behavior of the atom in an equation similar in many respects to the equation describing Maxwell's electromagnetic waves. As a result, Schroedinger's approach came to be called *wave mechanics*. The solutions to Schroedinger's equation contain an expression called the *wave function*, whose interpretation was to play a key role in attempts to describe the nature of the subatomic world. Since probability did not seem to enter into Schroedinger's expression, both Einstein and Planck hailed Schroedinger's work. Schroedinger seemed to restore the deterministic language that Bohr's probabilistic model of the atom had apparently ruled out.

Suddenly there was an embarrassment of riches. A year before there had been no adequate model of the atom, and now there were two! Heisenberg's matrix model, which made no attempt to describe atomic structure, and Schroedinger's wave model, which seemed to allow a classical wave interpretation of atomic structure. Which, if either, was correct? Here, as in the case of the work of Newton and Hamilton, the answer was provided not by arguing the merits of the two approaches but by demonstrating that the two theories are mathematically equivalent. In this case, Schroedinger provided the demonstration. Once again physicists took the most important aspect of a model to be its mathematical expression. Despite the fundamentally different philosophical approaches of Heisenberg and Schroedinger and the different ways they expressed their ideas, their models are held to be identical because the mathematical formulations of both approaches can be shown to be equivalent.[27]

Schroedinger wanted to identify the solutions to his equations as representing the "orbit" of an electron around an atomic nucleus. He developed several interpretations involving matter waves based on analogies with electromagnetic waves, but each of these attempts proved unsatisfactory. For example, a problem arose when physicists attempted to use Schroedinger's approach to predict the behavior of a free electron (one that is not bound to an atomic nucleus).

When energetic electrons from radioactive processes pass through matter such as a gas, they knock other electrons free from the atoms in the gas, creating charged atoms called *ions*. If the gas is made up of the right components, the ions become centers around which droplets form, leaving a track that traces the path of the energetic electron.

A cloud chamber is a device that can be used to study the behavior of energetic charged particles. Yet contrary to the well-defined path physicists see, when they used Schroedinger's equation to predict the behavior of free electrons, they found something quite different. The equation seemed to call for an electron that spreads out in space, like an expanding puff of smoke, as time goes on. How could Schroedinger's prediction be reconciled with cloud-chamber observations?

Determinism Dethroned

In June 1926, six months after Schroedinger's paper was published, the German physicist Max Born published a paper that struck at the heart of the conventional interpretation of physical reality. "It is necessary," he wrote, "to drop completely the physical pictures of Schroedinger which aim at a revitalization of the classical continuum theory, to retain only the formalism and to fill that with a new physical content."[28] Born maintained that the solutions to Schroedinger's equation should not be thought of as representing electrons at all. Rather they should be thought of as being related to the *probability* of finding an electron at any point in space.[29] Someone compared the Schroedinger wave to a crime wave. A crime wave is not a physical wave; it is an abstraction whose observable consequence is an increased probability that crimes will be committed.

Born's interpretation is much more radical than it may at first appear; it calls into question the most fundamental assumptions of classical physics. It abandons a rigid chain of Newtonian or Maxwellian determinism in favor of a fundamentally probabilistic view of reality. According to Born, Schroedinger's equation describes not the behavior of electrons but the probability of finding electrons at particular places.

If you flip a coin 100 times you can expect heads to come up roughly half the time. You can verify this conclusion by actually flipping a coin several thousand times and seeing how many times heads turn up in each run of 100 throws. The classical model says that every time you flip the coin there is a reason it comes up either heads or tails. Whether the coin comes up heads or tails has to do with the exact distribution of matter in the coin, the way you hold the coin, the exact amount of force you use and exactly where you apply the force to the coin, how the coin hits the floor, the roughness of the floor, and possibly even the air currents in the room. A good magician can consistently produce heads or tails, demonstrating that flipping a coin need not be a random process.

In the classical view, the fact that we can only predict the probable outcome of flipping a coin reflects our ignorance of these details, not anything fundamentally uncertain about the outcome of each coin flip. The same is true of mortality tables; given a person's age, weight, and sex, we can give the probability that he or she will live to be a certain age. Nevertheless, we believe everyone dies for a reason, even if we cannot predict what that reason will be; no one dies because of the "probability of death." In classical physics, randomness is only an illusion that arises in a world rigidly determined by physical laws.

People first said that the solutions to Schroedinger's equation (the wave function that provides a measure of the probability of finding an electron at any point in space) should be interpreted as being analogous to the mortality tables, that is, the wave function tells something about the limitations in physicists' *knowledge* of the electron's location. By taking this approach physicists could retain the classical way of talking, in which an electron has a definite location at every moment. They just did not happen to know what this location was. (In the same way, everyone will die for some specific reason, although we do not happen to know what it will be.) However, this interpretation has serious problems. We can see why by considering an experiment similar to Young's original experiment that displayed interference using light.

Young's experimental arrangement involved two narrow slits illuminated by a single light source. A screen placed behind the slits shows an alternating pattern of dark and light bands parallel to the slits. Young interpreted these bands as displaying the interference between light waves passing through each of the two slits.

A physicist can place a phosphorescent screen behind the two slits in Young's experiment and direct a stream of electrons instead of photons toward the slits.[30] A tiny flash of light then reveals where each electron strikes the screen. When this experiment is carried out, electrons are found to "prefer" locations in bands parallel to the slits and to "avoid" locations between these bands, just as photons do in Young's experiment. If Schroedinger's waves refer to knowledge of a quantum mechanical system, it appears we are being asked to believe that the patterns on the screen are the result of interference produced by "knowledge" passing through each of the two slits. But how can knowledge be said to pass anywhere?

Let us look at the experiment in a little more detail. Each electron arrives at the screen and produces a single flash. These flashes are small in size when compared with the overall distribution of flashes on the screen, so the electron appears to be small in comparison with the size of the apparatus and the slits. It seems reasonable that each electron has to pass through one slit or the other (if both slits are closed, no electrons are detected; if either slit is open, electrons reach the screen). The observed pattern should then be produced by the electrons that pass through the first slit plus the electrons that pass through the second slit. This deduction can be tested by opening one slit at a time and detecting the electrons that pass through each slit as they strike the screen. After the experiment has been run for each slit individually, the two sets of data can be added to find the combined effect of electrons passing through each of the two slits. When this two-stage experiment is performed, no interference pattern is found.

If both slits are open at the same time, however, an interference pattern clearly emerges. In fact, electrons seem to *avoid* destinations on the screen that are reached when only one slit is open at a time. Thus there is no way the interference pattern can build up by combining the patterns produced when only one of the two slits is open at any time. Apparently electrons passing through one slit "know" whether the other slit is open or closed and adjust their arrival on the screen accordingly. Does this behavior make sense?

Perhaps the interference is produced by the interaction of electrons passing through one slit with those passing through the other slit. We can test this possibility by lowering the intensity of the source until we calculate that on average only one electron is present in the apparatus at a time. The same interference pattern is still produced, so it cannot be the result of electrons interacting with each other in the apparatus.

Somehow, an electron passing through one slit seems to "know" whether the other slit is open. Each electron is detected as a single particle, and where the particle arrives on the screen seems to depend on the number of paths it *might* have taken: If both slits are open, the electron might have taken a path that carried it through either slit. The path the electron did not take apparently affects the path it did take. Recall that the French mathematician Pierre de Fermat described the behavior of light in the following way: The path light takes allows it to get from one place to another in the shortest possible time. Fermat's principle seems to suggest that light *knows* the possible paths and chooses the one that takes the least amount of time. How could light know beforehand which path would be the quickest? How could an electron know both slits were open?

7

The Intransigent Presence of Paradox

Quality is reduced to quantity: The number of electrons and the quantum numbers of a given state fully determine all properties of the atom in that state . . . the "harmonies of the spheres" reappear in the world of atoms. . . .[31]

VICTOR WEISSKOPF

Take a piece of paper and write on one side, "The sentence on the other side of this paper is true." Then turn the paper over and write, "The sentence on the other side of this paper is false." You have created a very peculiar situation. If the sentence you wrote first is true, then the second sentence tells you the first sentence must be false. In other words, if the first sentence is true, then it must be false. On the other hand, if the first sentence is false, then the second sentence must be false, and if it is false, the first sentence must be true.

This paradox arises because the two sentences, each of which by itself is perfectly reasonable, are not consistent when combined. Paradoxes like this one have bothered logicians since the time of the ancient Greeks. Physicists, however, were not used to dealing with this kind of distraction. Talk that combined the particle and wave descriptions of subatomic particles seemed to pose a similar paradox.

Even when physicists accepted Max Born's probabilistic interpretation, they still faced the problem of understanding how to use the apparently inconsistent ways of talking about atomic particles and light in the languages of waves and particles. One way of approaching the problem is through the relationship between the position of a particle and its motion—a relationship Heisenberg uncovered when he developed his way of describing the emissions from atoms.

Heisenberg's relationship, sometimes called the *uncertainty principle,* says that physicists can determine an electron's position to any precision they like, but the more precisely they determine this position, the less precisely can they determine how the electron is moving at the same moment. In other words, they can find out exactly where the electron is, but only at the price of being able to say nothing about where it is going. On the other hand, physicists can measure an electron's momentum as precisely as they like, but the more precisely they know the momentum, the less precisely they can tell simultaneously the electron's position. They can tell exactly where the electron is going, but only at the price of being unable to say anything about where it is.

Imagine a device that shoots marbles and a screen that flashes to show where a marble strikes it. Now imagine that a plate with a hole larger than the marbles is placed between the source of marbles and the screen. The plate serves to reject some marbles, and the hole allows other marbles to pass through and strike the screen. Where the marbles strike the screen depends on how big the hole is. We could minimize the size of the region where the marbles arrive, and thus predict their paths more precisely, by making the hole smaller, as long as we do not make the hole so small that a marble could not pass through it.

Now imagine a similar experiment with electrons. When we start out, the electrons seem to behave just the way the marbles do—we can locate the arrival of each electron as it strikes the screen and thus get a rough idea of the electron's path. As we make the hole smaller, something quite different happens: Instead of being confined to a smaller and smaller area, the flashes indicating the arrival of an electron scatter more and more widely over the screen. If we make the hole small enough, the electrons can arrive almost anywhere on the screen. Making the hole smaller makes it *easier* to predict where the *marbles* will arrive at the screen. Making the hole smaller makes it *harder* to predict where the *electrons* will arrive. After the electrons pass though a very precise position localizer (the hole in the plate), we can say almost nothing about what the momentum, and hence the subsequent path, of the electrons will be.

The behavior of electrons can also be discussed in the vocabulary of waves. You may recall that de Broglie postulated a relationship between an electron's momentum and its wavelength or frequency. De Broglie's argument implies that the frequency associated with an electron must be known very precisely if the momentum of the electron is to be known with comparable precision. A long "train" of uniform waves is needed to allow a precise measurement of the frequency to be made.[32] A long train of waves, however, has no well-defined "position." But there is a way to describe the position of an electron in the wave vocabulary. If waves with different wavelengths are combined, they will sometimes cancel and sometimes reinforce each other, just as Young said the waves behaved in his nineteenth-century two-slit experiment using light. By very carefully choosing many waves with different wavelengths it is possible to arrange for the waves to cancel each other out everywhere except in the one region where the electron is located. Such a collection of waves, however, has no well-defined momentum, since each wave in the collection has a different momentum.

The wavelike nature of electrons makes it impossible to say that they follow well-defined paths through space and time. On the other hand, the particlelike nature of electrons means that physicists can say that electrons have a very specific mass and electric charge. The combination of these two descriptions does not seem to resemble anything we can even picture, much less experience. Physicists cannot give up either side of this paradox, however. If electrons were not particles, it would be impossible to explain how a television set works, but if these particles had well-defined paths, there would be no way to explain the interference seen in the two-slit experiment.

The problem with trying to understand the behavior of electrons arises, Heisenberg said, because we persist in thinking of electrons as tiny marbles; we persist in talking as if there were subatomic "objects" that physical theories somehow describe. But electrons are not objects in this sense at all. Heisenberg maintained quantum theory tells us *everything* we can expect to find out about the world as far as the "behavior" of the systems encompassed by the theory is concerned. Asking what the behavior of electrons is "really" like arises out of the marble fallacy. Such questioning is futile. At best any answer is simply a matter of taste. Discussions that do not lead to any new predictions have no impact on science; discussions that lead to new predictions are challenges to be met by experiments in the laboratory.

The histories of electrons, like any histories, are stories people tell about the world. The paths of electrons are something physicists are free to make up, because they can never test their conjectures. Thus they can *say* that the path of an electron in their experimental setup was anything they like. (Physicists normally say the path of an electron is a straight line if no magnetic or electric fields are present, because this makes the description of the cathode-ray tube, for example, simple. We will see in the next chapter that it is possible to talk about this path in very different ways.)

When an astrophysicist shows us the photographic image of a distant galaxy, she tells us the photons that formed the image left the galaxy, possibly several billion years ago, traveled on more or less straight paths through space, and reached a detector attached to a telescope. At the telescope, the arrival of the photons was recorded and an image built up. Heisenberg tells us that the astronomer's conviction is a just-so story. There is no way it could possibly be tested. Granted, it is a carefully constructed just-so story—its consistency with other just-so stories can be rigorously checked—but it remains a matter of taste as far as experimental physics is concerned. Only when physicists talk about what has yet to occur can their statements be tested.

After many attempts to show otherwise, it has become clear that no experiment can be developed that allows the path of an electron to be determined precisely. Scientists had faced unanswered questions many times in the past, but they always had faith that future information would bring understanding. Newton, for example, was convinced that it would be possible eventually to understand the nature of gravity. Heisenberg, however, held out no such hope about any future understanding of the paths of atomic particles. From his perspective, physicists had encountered an absolute limit to what they can say about nature.

Even if we do not observe a marble, we can still say that it has a well-defined path. We can say this because we can use the path to predict outcomes that can be observed. Electrons are not like marbles. When we do not measure an electron's path, we cannot even say that it *has* a path. Position and motion do not seem to be properties of the subatomic world; they seem to be our way of *talking* about the subatomic world.

For those who insist on trying to picture the path of photons or electrons despite Heisenberg's warnings, the English Nobel laureate Paul Dirac has the following words:

The main object of physical science is not the provision of pictures, but is the formulation of laws governing phenomena and the application of these laws to the discovery of new phenomena. If a picture exists, so much the better; but whether a picture exists or not is a matter of only secondary importance. In the case of atomic phenomena no picture can be expected to exist in the usual sense of the word 'picture,' by which is meant a model functioning essentially on classical lines. One may extend the meaning of the word 'picture' to include any way of looking at the fundamental laws which makes their self-consistency obvious. With this extension, one may acquire a picture of atomic phenomena by becoming familiar with the laws of quantum theory.[33]

Classical physics replaced the question *why* with the question *how*. In the languages of Newton and Maxwell, answers to the latter question come in the form of descriptions of motion through space and time. The new language of quantum mechanics replaces the question *how* with the question *what*. What is the outcome of carrying out an experiment? Quantum mechanics is not a picture or a narrative that tells physicists *how* the outcome of an experiment comes about; it is a mathematical expression that allows them to calculate *what* the outcome of an experiment will be.

Niels Bohr gave a great deal of thought to the implication and meaning of quantum mechanics. He concluded that we must take seriously the fact that physics is our way of talking about the world. Bohr said, "There is no quantum world. There is only an abstract quantum physical description. It is wrong to think that the task of physics is to find out how nature *is*. Physics concerns only what we can *say* about nature."[34]

Physical theories do not tell physicists how the world *is;* they tell physicists what they can predict reliably about the behavior of the world. Bohr reminded physicists that physical theories are models in the Ptolemaic sense—ways of making calculations—and not pictures of a world that is somehow independent of their descriptions.

The wave function that forms the solution to Schroedinger's equation does not picture some*thing* in nature. Physical models of the subatomic world relate to the world, not on a point by point basis, but in terms of destinations. Quantum mechanics can be used to predict the probability of finding electrons somewhere but not the paths they will take to arrive there; it provides a description of *what* happens but gives no detailed description of *how* this happens. The solutions to Schroedinger's equation resemble airline schedules more than they resemble maps: They tell us about possible destinations without providing a description of the routes traveled to reach these destinations.

Since the time of Newton, if not before, physics has been thought of as the search for the laws of nature—the laws describing how God established the natural world. From Bohr's perspective this search is misdirected; the laws of physics are *our* laws, not nature's.

Not being able to say a particle has a precise path seems to some people to be equivalent to saying that the particle is not "real." To such people modern physics seems to be saying that the world is the subjective creation of individual observers. But there is nothing subjective about the methods of physics; the flash on the screen that heralds the arrival of an electron is as objective as anything can be. It is true that some things physicists thought made perfect sense, like the paths of electrons, do not turn out to be useful ways of talking, but this discovery does not make the world any less objective. The feeling we have that this realization makes the world "subjective" only shows how wedded we are to the idea that what we say pictures an "objective" reality that is independent of the way we interact with it. But this picture theory of language is a metaphor—a metaphor we are so enamored of that being asked to give it up seems like being asked to give up reality itself.

Let us return to the example of the mortality tables used by life insurance companies. Do these tables provide a picture of some aspect of reality? Most of us do not believe there is some inviolable principle of mortality that demands that a certain number of people die in automobile accidents every year. Rather, we think we can guess how many people will die based on what we know about the past history of automobile accidents. The mortality tables are a way of talking about our experience, yet they do not picture reality. Bohr tells us physics is also a way of making predictions. The solutions to Schroedinger's equation resemble the mortality tables more than they resemble photographs. We can never test the belief that the world must be some particular way whether we observe it or not. The paths of electrons are like elves that cavort in the forest only when there is no one there to see them. Anyone who likes can believe in the existence of such elves, but this belief should not be confused with physics.

If we were to take Schroedinger's equation to be a description of the behavior of the world, rather than a device for calculating probabilities, we would be forced to say something like the following: "An electron starts out as an abstract mathematical wave that passes through the two slits and arrives at the screen after interfering with itself. This abstract wave then instantaneously collapses and is replaced by an electron when the wave arrives at the screen."

The abstract wave must collapse instantaneously because the instant an electron is detected there is zero probability of detecting the same electron somewhere else. This instantaneous collapse seems to involve speeds faster than light (no matter how large it is, the entire wave must "know" instantaneously that an electron has been detected; otherwise one electron might be detected at two different places at the same time). But interestingly enough, physicists have found no way to use this instantaneous collapse to send a signal faster than light and thereby to violate special relativity. Still, the idea of instantaneous "communication" over great distances leaves some physicists very uncomfortable.

Whenever a flash on the screen occurs, it shows that an electron has been detected. The flash is always located at a well-defined place, so physicists say the electron is located there too. Whenever physicists measure the mass and electric charge of an electron, they arrive at the same well-determined values. In all these respects the electron seems to be a perfectly self-respecting particle. Whenever physicists do not observe an electron, however, they cannot pinpoint its location and motion, which must be, for example, sufficiently ambiguous to allow the electron to pass through either of the two slits in the experiment.

Physicists can calculate the probability of detecting the electron at any point on the screen. They can say nothing about exactly how the electron got there, or what its path must be like when they are not measuring it, because to test the validity of their description, they would have to set up a different experimental arrangement altogether, and a different arrangement would lead to a very different outcome. In fact, whenever physicists set up an apparatus capable of telling them which slit the electron passes through, they destroy the interference phenomenon. Interference exists only so long as it is impossible to tell what path the electron follows. In the words of Richard Feynman:

There was a time when the newspapers said that only twelve men understood the theory of relativity. I do not believe there ever was such a time. There might have been a time when only one man did, because he was the only guy who caught on, before he wrote his paper. But after people read the paper a lot of people understood the theory of relativity in some way or other, certainly more than twelve. On the other hand, I think I can safely say that no one understands quantum mechanics. . . . Do not keep saying to yourself, if you possibly can avoid it, 'But how can it be like that?' because you will get 'down the drain', into a blind alley from which nobody has yet escaped. Nobody knows how it can be like that.[35]

We may think we are making sense when we talk about what the world is doing whether we observe it or not. It seems perfectly reasonable, for example, to say there is sound in the forest when a tree falls, whether or not there is anyone around to hear it. Quantum mechanics gives no support to this notion. The world on the atomic scale, at least, does not seem to be some particular way, whether physicists observe it or not. The atomic world appears to have particular qualities only as the result of measurements physicists make. Quantum mechanics is a way of talking about nature that allows physicists to predict how the world will respond to being measured. So long as we stick to this understanding, quantum mechanics raises no problems. If, on the other hand, we persist in demanding to know how the world *is,* independent of how it appears to be in experiments, we, in Feynman's words, "will get 'down the drain', into a blind alley from which nobody has yet escaped."

We firmly believe that we relate to a world with "objective" characteristics that somehow exist independently of what we observe or what our theories tell us. Giving up this belief may seem to be a high price to pay, but it is the price physicists have had to pay in exchange for being able to predict the outcome of their experiments.

8

The Inexhaustible Fecundity of Space

Dick Feynman told me about his "sum over histories" version of quantum mechanics. "The electron does anything it likes," he said. "It goes in any direction at any speed, forward or backward in time, however it likes, and then you add up the amplitudes and it gives you the wave function." I said to him, "You're crazy." But he wasn't.[36]

FREEMAN DYSON

The 50 years following are the story of the difficult birth of a language that unifies quantum mechanics and relativity, the language called *quantum field theory*. Quantum field theory had its beginnings in the 1920s, but it was not until the 1970s that physicists became convinced that this way of talking about nature was sufficiently useful and powerful to win a central place in the science.

An Infinite Sea

In 1927 an English physicist, Paul Dirac, wrote down an equation that for the first time successfully unified quantum mechanics and relativity. The equation was immediately successful in allowing physicists to deal with a variety of problems whose solution had eluded them up to this time. But Dirac's equation raised an immediate problem—it had two types of solutions, and one of them called for electrons with negative energy. Now if the idea of energy is abstract, the idea of negative energy (and by virtue of special relativity, negative mass) is bizarre. For example, in order to have negative kinetic energy, the cars on our roller coaster would have to move at imaginary speeds or have negative mass. Physicists had no idea what negative energy could mean.

Bohr said that the atom is stable because there is a lowest energy associated with each atom—the so-called *ground state*. An electron in a hydrogen atom, for example, can give up energy only until it reaches the ground state. If Dirac's equation were correct, there would be an infinite number of negative energy states having even lower energy than the ground state. There would seem to be nothing to prevent all

103

the electrons in ordinary matter from loosing an unlimited amount of energy by dropping into these negative energy states. Matter should be unstable—it should vanish in the wink of an eye. Clearly something was wrong.

Dirac identified the positive energy states with electrons, and since he could find no way to eliminate the negative energy solutions to his equation, he decided that the mathematics required that he accept the existence of negative energy electrons. Dirac's solution to the problem of negative energy was to talk in terms of a vacuum that, instead of being empty, is filled by an infinite number of electrons with negative energy. Dirac said the world as we know it is floating on a sea of negative energy electrons. At first, posing the existence of an infinite number of electrons may seem like a rather extravagant solution to the problem, but Dirac argued that there was no way to detect directly these negative-energy electrons.

When a bird stands on a high-voltage line, the bird does not experience a shock because it is at the same electrical "potential" as the power line. All that a bird, or a human being for that matter, is sensitive to is a *difference* in voltage. Only when the bird simultaneously touches both the electric power line and a tree, for example, does the bird create this difference, which in this case is likely to prove fatal. The negative-energy electrons do not make us aware of their presence in quite so dramatic a fashion, but just as the high voltage on the power line is normally invisible to the bird, Dirac's infinite sea of negative-energy electrons is normally invisible to us.

In Dirac's description there is an analogy with the bird touching the wire and the tree at the same time—a process by which a normally invisible negative-energy electron becomes observable. Occasionally, an energetic photon will strike a negative-energy electron and add enough energy to make the electron's energy positive. An electron with positive energy is simply an ordinary electron. Dirac said that this process would leave behind a "hole" in the negative-energy sea. He asked what this hole might look like; one clue was the behavior of electric charge.

We have looked in some detail at the conservation of energy in our roller-coaster example. A somewhat different kind of conservation is the *conservation of electric charge*. Electric charge comes in discrete units—there are no fractional electric charges in nature. Physicists found that if they added up the positive charges and subtracted the negative charges for any combination of particles *before* the particles underwent any interaction, and did the same thing *after* the interaction, they always got the same number. This property of nature is called the conservation of electric charge. No violation of charge conservation has ever been observed.

It would seem that when a negative-energy electron absorbed enough energy from a photon to become a normal electron, an isolated negative charge would suddenly be created. Such an occurrence would violate the conservation of charge. In order to preserve this idea, Dirac proposed that the "hole" left in the negative-energy sea appears to us to be a positively charged particle. Only one particle with a positive charge was known at that time, the proton. Dirac, therefore, associated protons with the positively charged particles his model seemed to require. Protons have almost 2,000 times the mass of electrons. Where could all this mass come from?

Newton's laws of motion can be reversed and still describe possible physical situations—if the direction of the motion of all the planets were reversed, they would still follow Newton's laws. In the same way, the reverse of Dirac's "hole" creation might also be a legitimate physical process. That is, a hole should be able to combine with a normal electron and emit a high-energy photon. But if Dirac's holes were protons, matter would be unstable—electrons could combine with protons and the two particles would disappear with a burst of energy. Interpreted in this way, the theory seemed to say that all matter should vanish in a fraction of a second! Dirac was discouraged, but eventually he was forced to conclude that accepting his way of talking about nature called for accepting the existence of a new fundamental constituent of matter: a positively charged electron. Although such a particle had never been seen, it should be observable.

Meanwhile, the American physicist Carl Anderson was using cloud chambers to study the extremely energetic particles constantly raining down on the earth from outer space—cosmic rays. Heisenberg seemed to tell us that atomic particles do not have well-defined paths through space and time; how then can physicists photograph the paths of atomic particles in a cloud chamber? The path we can see in

105

a cloud chamber is not subatomic—the trail is far larger. As droplets condense on the ions left by the passage of the cosmic-ray particle, a trail is formed that is roughly a million times larger than the particle producing it. The trail is so large, in fact, that the consequences of the indeterminacy relations are hidden from us inside the volume of the trail and prevent us from being able to say the particle has a definite path within the trail.

Strange events were sometimes revealed in the cloud chambers that physicists used to study cosmic rays. The most common tracks in cloud chambers were found to be produced by electrons. Occasionally a track would appear that looked just like the track of an electron but seemed to bend the wrong way in the magnetic fields used to study the charge of cosmic rays. The tracks might seem to bend the wrong way if they were produced by electrons moving in the "wrong" way—moving upward from the earth rather than downward from space. It was also possible that the strange tracks were produced by protons, but the tracks of protons are denser than the tracks of electrons, and the suspicious tracks looked exactly like those of electrons.

By placing a lead sheet in the cloud chamber, Anderson slowed down the cosmic-ray particles. Slower particles curve more strongly in a magnetic field, and Anderson found that the particles' paths curved just as they would be expected to if the tracks indeed were made by particles coming from space. He published his result and tentatively called for the existence of a new kind of particle, the *positron*. Positrons are identical to electrons except they carry a positive, rather than a negative, electric charge. Without knowing it, Anderson had vindicated Dirac; although Dirac did not at first realize it, the language he created required the existence of a new fundamental constituent of nature—antimatter. It would take until 1955 for physicists to create antiprotons and complete the demonstration that antimatter can exist, but the discovery of the positron convinced physicists of the power of Dirac's language. The existence of antimatter means that physicists had to give up talking as though matter is immutable. Dirac argued that particles can disappear in a burst of energy (when a positron encounters an electron, for example) or be created out of energy (when a high-energy photon converts itself into an electron and a positron).

Despite the predictive power of Dirac's equation, many physicists were very unhappy with his interpretation of what the equation entailed. They found it very hard to accept the idea of an infinite sea of negative energy. In 1934 the American physicists Robert Oppenheimer and Wendell Furry showed that Dirac's equation *could* be interpreted in another way. In the negative-energy sea description, a photon can create an electron-positron pair by providing enough energy to a negative-energy electron. By redefining terms, Oppenheimer and Furry were able to eliminate talk of an infinite sea of negative-energy electrons and replace it with talk of the creation and destruction of electrons and positrons. The mathematical theory remained the same; only the way physicists talked about the theory changed. This change may have comforted those who disliked Dirac's infinite sea, but other problems remained—problems that seemed insurmountable.

Filling the Vacuum

Imagine that an electron briefly winks into, and then out of, existence. Einstein allows us to say there is energy associated with the mass of this electron, but time is needed to measure this energy precisely. The same Heisenberg relationship that tells physicists that they cannot precisely and simultaneously measure both the position and motion of an electron tells them there is a limit to the precision with which they can know simultaneously both energy and time. Physicists can measure energy as precisely as they like, but they must take time to carry out the measurement—the more precisely they determine the energy, the longer the time required. If the interval during which an electron exists is very short, physicists cannot measure its energy accurately; in other words, they cannot even be sure the electron is there at all. Thus physicists can talk about processes that violate the conservation of energy by creating matter and, hence, energy, out of the vacuum but only if they talk about processes that take place in so short an interval that the violation can never be measured. If they cover their bad checks before they can be cashed, no one will be the wiser.

The equations of quantum field theory describe fields that harbor the potential for the continuous creation and destruction of the particles associated with the field. To distinguish these particles from particles that persist in time, the evanescent particles are called *virtual,* as opposed to *real* particles. In addition to virtual electrons and positrons, physicists also talk about virtual photons.

The calculations required to get accurate answers based on Dirac's equation take the form of successive corrections to initial approximations. The initial approximations are reasonably accurate, but in order to improve the precision of an answer, it is necessary to refine the calculations. This step, however, introduces real headaches because of the virtual processes said to be continuously going on in the vacuum. An electron should interact with virtual electrons and positrons, as well as with virtual photons that the electron itself produces. The problem is that these interactions, which physicists hoped would lead to small corrections to the results from the initial approximations, turned out to contribute infinitely large corrections. One infinity arises because an electron can emit a virtual photon and reabsorb the same photon after a vanishingly short period of time. Thus, according to Heisenberg's relationship, the virtual photon could have an infinite amount of energy and this energy would contribute to the energy, and hence mass, of the electron. The mass of an electron, however, is well determined and is obviously nowhere near infinitely large. With infinities popping up in the calculations, there seemed to be no way to use quantum field theory to provide accurate descriptions of the physical world.

At times the problem seemed so intractable that many physicists, including Dirac, thought that the entire language of quantum field theory would have to be abandoned—a radically new way of talking about the subatomic world seemed to be needed. But in this case the examples of the radically new vocabularies of relativity and quantum mechanics were misleading. Major revisions were not necessary—physicists were closer to a solution than they thought, although almost 20 years passed before they discovered how close they were.

At first the problem of infinities was more troubling as an idea than as a practical limitation. The Austrian physicist Wolfgang Pauli said, "Success seems to have been on the side of Dirac rather than of logic."[37] Solutions to Dirac's equation that only involved the lowest order of approximation seemed to describe what physicists found in their laboratories with adequate precision. Apparently the refined corrections and their associated infinities could be safely ignored.

The breakthrough in dealing with the problem of infinities came in the years immediately after World War II, and was spurred by laboratory measurement. In 1947 the American physicist Willis Lamb made a very precise measurement of the difference between two energies that an electron could have in a hydrogen atom. This measurement was clearly discordant with calculations of the difference made using Dirac's equation. Another measurement that conflicted with Dirac's theory was made at almost the same time. One of the early triumphs of the Dirac theory was its ability to describe accurately the strength of the electron's magnetic field; however, more refined studies by Isidor Rabi, John Nafe, and Edward Nelson showed a small but clear discrepancy between the observed and calculated values.

The difficulty of dealing with the infinities that arise in the interaction between the electron and the electromagnetic field led physicists to ignore these corrections, hoping somehow they would not affect the final answer. The new experimental results indicated that the interactions must make a small contribution—so while the terms could not be ignored, somehow there had to be a way to eliminate the undesirable consequences of the infinities.

New Definitions

The freezing point of water is located at different temperatures on the Fahrenheit and Centigrade scales, and the size of a degree is different on the two scales. But water freezes when water freezes, and it seems reasonable that our description of nature should not depend on the scale we use to measure temperature—just as a description of the weather should not depend on whether it is given in French or English. Temperature scales, like all scales, are a human convention.

The solution to the problem of infinities came with the development of a variety of mathematical techniques—some would say tricks—that allow positive infinities to cancel negative infinities. This cancellation amounts to a redefinition of the physical mass and charge—a new convention. The "new" mass and electric charge are set equal to the observed mass and charge, and the calculation proceeds. As a result of this process, physicists cannot calculate the mass and charge of the electron from first principles; the values they use must be measured in the laboratory and entered into their equations.

Allowing positive infinities to cancel negative infinities is not legitimate as far as mathematicians are concerned, but physicists had no other alternative and hoped these problems would not prove insurmountable. When the calculus was first developed independently by Newton and Leibniz, many mathematical techniques were used successfully for years before mathematicians were able to develop a rationale to justify the fundamental concepts. Since infinities were involved in this case too, there was at least a precedent for not worrying too much about the problem.

The American physicists Julian Schwinger and Richard Feynman played a key role in developing the details of this redefinition. Feynman took a different tack from that of most physicists. Initially, he had decided that rather than being the solution, the field was the problem. He hoped it would be possible to avoid the problem of the infinite energy created by the electron's interaction with its own field by adopting a vocabulary in which there is no such interaction. This hope did not ultimately work out, but the approach led Feynman to revive Newton's way of talking about forces.

Feynman explored what would happen if he talked in terms of electric forces that act directly over distances without any intervening field. Feynman's electromagnetic action-at-a-distance differs from Newton's gravitational action-at-a-distance in that Feynman found that he had to make use of electromagnetic effects that travel both forward and backward in time. As strange as this notion appears, he was able to show that this description leads to exactly the same description of the world that Maxwell's equations lead to. The essential point is not the concept of a field but the equations that allow physicists to calculate what electrons will do in different circumstances.

In his approach to understanding the interaction of electrons and the electromagnetic field, Feynman considered the entire history of the electron from the time it left the source to the time it arrived at the detector. But didn't Heisenberg convince us that the electron cannot be said to have a path in space and time? Feynman got around this problem by considering the *possible* paths an electron might take in moving from one point to another. Feynman gives a different explanation from Heisenberg's for why it is impossible to ascribe a unique path for the electron. Feynman says the probability of finding an electron at any point on the screen depends on considering the probabilities associated with the routes the electron *might* have taken to get there, no matter how unlikely each route might be. This way of talking is closely related to Hamilton's assertion, which we encountered earlier, that an object somehow knows enough to take the route that minimizes a quantity, called the action, associated with its energy.

An electron arrives at a destination in a way that minimizes the action associated with its trip. How does the electron know which paths it might have taken and what action is associated with each of these paths? Apparently the same way the electron knows in the two-slit experiment that both slits, not just one, are open. Talking about electrons *as if* they knew how to minimize the action leads to useful predictions.

Feynman developed another way of talking about the subatomic world. His approach leads to different techniques for making calculations, but his language leads to exactly the same conclusions about what will be observed as does the "traditional" view. Hence physicists judge the two approaches to be equivalent.

In trying to simplify the calculations involved in predicting what will happen when electrons and light interact, Feynman developed a way of diagramming the many interactions involved. Physicists had invented ways to describe the creation and annihilation of positrons and electrons, but Feynman found what for him was a simpler way— he treated positrons as electrons moving backward in time. This technique served to ensure that the processes Feynman considered are consistent with relativity. Are positrons "really" electrons moving backward in time? This question is resolved the way so many similar questions in physics are resolved—by saying that the same results are obtained by treating positrons as electrons moving backward in time as are obtained by using more conventional techniques, and so the ideas are equivalent. In other words, as far as physics is concerned, whether positrons really are electrons moving backward in time is not a useful question.

Feynman diagrams are used constantly by physicists, but when he first introduced them, the idea was so revolutionary that Bohr was sure Feynman didn't understand elementary quantum mechanics![38] Bohr's consternation was produced by Feynman's depiction of the paths of electrons as straight lines—exactly the sort of precise paths that Heisenberg's relationship ruled out. Feynman tried to explain that the diagrams were bookkeeping devices and not meant to be pictures, but this distinction eluded Bohr. The grip of the idea that the things we say, our models, picture nature, is so strong that Niels Bohr, the man who most clearly saw that physics is a way of talking, failed at first to realize that Feynman had developed a new and immensely productive way of discussing nature.

QED

Feynman diagrams exemplify one problem with understanding the interactions between electrons and the electromagnetic field. We are comfortable with what we can visualize. Feynman developed the diagrams as a way of keeping track of the complex steps involved in the calculations, but it is sometimes difficult to avoid taking the step of identifying a visual representation as a picture and to avoid making Bohr's mistake. A graph may *represent* the population of the United States, but the graph doesn't *picture* the population; we can visualize both the graph and the population and determine that the graph does not resemble 240 million people.

If a graph depicts the growth in the population of the United States from 1620 to 1980, we are unlikely to think of it as a picture even though we cannot visualize the growth and compare it with the graph. The growth is something we can represent but not something we can photograph—it is more of an abstract idea than an object, and we know we cannot make a picture of an abstract idea. Electrons, on the other hand, are not supposed to be abstract ideas, are they? Instead they are supposed to be things, and it is very hard to resist the desire to picture a thing. We have to be very careful about giving into this desire, lest, in Feynman's words, we "get 'down the drain', into a blind alley from which nobody has yet escaped." There are times when we are better off thinking of electrons as abstract ideas—as ways of talking about nature.

Feynman developed a variety of shortcuts and assured himself that they were legitimate by comparing his results with those based on more orthodox methods. More concerned with arriving at answers than with mathematical niceties, he developed a variety of simplified ways of making calculations in which the infinities that had blocked physicists for the preceding 20 years could be eliminated by absorbing them into a redefinition of the mass and charge of an electron.

Schwinger developed his own approach to eliminating infinities, a method of considerably greater mathematical sophistication than Feynman's. However, when the two men compared their answers, they found they agreed. Meanwhile, working in total isolation in Japan during the Second World War, Shinichiro Tomonaga arrived at yet another way of eliminating infinities that also produced the same answers.

After 20 years of drought, the embarrassment of riches of three successful formulations of *quantum electrodynamics,* or *QED* as the language is called, reminds us of the independent development of quantum mechanics by Schroedinger and Heisenberg. In each case the denouement was similar; in 1949 the English physicist Freeman Dyson demonstrated that the three methods are mathematically equivalent. Once again physicists took apparently widely disparate theories to be equivalent because their underlying mathematical structure was identical.

Not everyone is thrilled by the process of eliminating infinities called *renormalization.* Some feel that it lacks the elegance and simplicity of either classical physics or quantum mechanics. In his Nobel acceptance speech Feynman said, "I think that renormalization theory is simply a way to sweep the difficulties of . . . electrodynamics under the rug."[39] More recently he called renormalization "a dippy process."[40] Dirac never accepted the process as anything other than a trick.

Nevertheless, renormalization has been so successful that almost all physicists admit that the technique is probably here to stay. Its most enthusiastic supporters, including American Nobel laureate Steven Weinberg, look at renormalizability as the hallmark of a successful physical theory: "But here was a principle—the infinities had to cancel [each other]—that was a golden key that explained why the theory was the way it was."[41] In other words, renormalizability is a clue that physicists have found a way of talking about nature that will turn out to be valuable.

QED demonstrates the ruthless stripping away of complications that physicists engage in to reach a situation simple enough to describe in mathematical language. In Feynman's formulation of QED only three fundamental things can happen: (1) a photon can go from one place to another; or (2) an electron can go from one place to another; or (3) an electron can emit or absorb a photon. Each of these has a probability associated with it, and physicists are concerned with calculating the probability associated with a particular outcome based on the many possible ways to arrive at that outcome from a particular starting point or, as a physicist would say, a set of initial conditions.

The Hungarian-born American Nobel laureate Eugene Wigner described the importance of initial conditions in this way:

> The world is very complicated and it is clearly impossible for the human mind to understand it completely. Man has therefore devised an artifice which permits the complicated nature of the world to be blamed on something which is called accidental and thus permits him to abstract a domain in which simple laws can be found. The complications are called initial conditions; the domain of regularities, laws of nature. Unnatural as such a division of the world's structure may appear from a very detached point of view . . . [it] is probably one of the most fruitful ones the human mind has made. It has made the natural sciences possible."[42]

Despite the doubts of its developers and the mathematically incomplete foundations for some of its calculations, QED is one of the most accurate descriptions of nature ever devised. Calculations of the values of fundamental physical constants such as the strength of the magnetic field of the electron agree with the best observations to one part in a hundred million—eight decimal places! (But recall that because of renormalization, the charge and mass of the electron cannot be calculated but must be obtained from measurement.) Not only is the theory accurate, but it also encompasses virtually all the processes taking place in the world around us. All chemical and biological processes are fundamentally quantum electrodynamical in nature. QED cannot, however, be used to predict the outcome of even the simplest biological processes. Biology is impossibly complex on the scale in which QED calculations are made. Someone who had just learned the musical scales would hardly be expected to compose a symphony simply because he or she knew what the ingredients were.

As a result of the great success of QED, physicists accept the continuous creation of particle-antiparticle pairs in empty space not only as a possibility but also as an actuality. Physicists talk about the virtual particles permitted by the Heisenberg relationships—particles that they can never observe—in the same way as they talk about particles they can observe. We can think of these omnipresent virtual particle-antiparticle pairs in the same way Dirac talked about negative-energy electrons; the pairs become visible only when something hits them with enough energy to "promote" them from virtual particles to real particles. The idea of empty space has certainly come a long way from the time when *vacuum* meant an absence of everything, a void. Perhaps Aristotle was not so far off the mark when he said that nature abhors a vacuum—at least human nature seems to abhor an old-fashioned vacuum with nothing in it.

Despite its great predictive power, the language of QED has not replaced Maxwell's electrodynamics any more than general relativity replaced Newtonian physics. Maxwell's equations remain as useful as ever, and for many problems his electromagnetic field is just as valuable a way of talking about the world as it ever was. Whether physicists talk in terms of classical field theory or quantum field theory depends on the problem they are wrestling with, much as the choice of a file or a saw depends on what a carpenter is trying to do.

In the language of Maxwell's theory, electromagnetism is a field—a collection of numbers assigned to every point throughout a region of space and time. Quantum field theory associates this field with the probability of the creation and destruction of a particle, the photon. Just as we can associate a particle with a field, we can associate a field with a particle, as de Broglie demonstrated. QED associates a field with the electron; the electron field is a way of describing the probability of finding an electron somewhere in a region of space and time. Electrons are described as manifestations of the electron field just as photons are described as manifestations of the electromagnetic field. The electron field is as abstract as the electromagnetic field—and as real.

The electron Thomson discovered in 1897 has become something he would surely never recognize. First, de Broglie showed the electron could be talked about in the vocabulary of waves as well as the vocabulary of particles. Next, Schroedinger's language (despite his reluctance to accept the interpretation) described the world in terms, not of the behavior of electrons, but of the probability of finding electrons in certain regions. When the electromagnetic field was included in this description, Dirac first interpreted the new language in terms of electrons and an infinite sea of negative-energy electrons with occasional "holes" that manifest themselves as positrons. Today electrons and positrons are described not as fundamental entities but as manifestations of a quantum field.

Perhaps we should not be too surprised with the way things turned out. Einstein gave us a glimpse of the peculiar nature of the stuff out of which the world is made when he showed us that it is necessary to talk about matter in the same way we talk about energy. From here on in it will be increasingly important to keep Bohr's admonition in mind: "There is no quantum world. There is only an abstract quantum physical description. It is wrong to think that the task of physics is to find out how nature is. Physics concerns only what we can *say* about nature."

The electron field apparently contains only one blueprint for the electron, because all electrons, insofar as anyone can tell, are absolutely identical. The Austrian-born American physicist Victor Weisskopf said:

> Within the framework of classical physics, it is hard to understand why there should not exist electrons with slightly less charge, or with a different mass, or with a spin (rotation about an axis) somewhat at variance with the spin of the observed electron. It is the existence of well-defined specific qualities, of which nature abounds, that runs counter to the spirit of classical physics.[43]

In a way, electrons share the property of indistinguishability with other abstractions such as numbers. Just as there are no differences between versions of the number two, there are no differences between electrons. The uniformity of electrons is a product of the way physicists talk about nature. Electrons have no properties other than their identical mass, electric charge, and so on, because the language of quantum electrodynamics requires that all electrons be identical. Something about nature makes it possible to talk about the world in terms of identical electrons. We will return later to the question of what this something might be.

The same inability that prevents physicists from talking about differences between individual electrons or the simultaneous position and motion of electrons give them the power to talk about the stability of the world—for if electrons were not identical, or if we could talk about the definite path of an electron, we would once again be faced with the inability of classical physics to explain how there can be any stable world at the subatomic level. Is it any wonder that despite its apparent paradoxes, the language of quantum mechanics is the heart of physical theory?

9

The Improbable Prevalence of Symmetry

Part of the art and skill of the engineer and of the experimental physicist is to create conditions in which certain events are sure to occur.[44]

EUGENE WIGNER

Most of us receive a statement every month that reveals our bank's commitment to a conservation principle. Although my bank does not care to whom I wrote my checks, and has absolutely no knowledge of why I wrote most of my checks, it is convinced of one thing: The balance in my account is *always* equal to the balance at the start of the month, plus the deposits I made, minus any service charges and minus the checks I wrote. A physicist would say that money is *conserved* in my checking account. (Money is not conserved in general, because governments and banks can "create" money with printing presses and loans—but that is another story altogether.)

The Unmeasurable Lightness of Neutrinos

Physicists are very fond of conservation laws. Conservation laws make it possible for them to say what things happen and what things do not happen without requiring them to know any details of *how* a particular process works. The story of neutrinos demonstrates the importance physicists place on conservation laws. The existence of the neutrino was first suggested in 1933 by the Austrian physicist Wolfgang Pauli. At that time, certain radioactive processes seemed to violate the conservation of energy. Pauli realized it would still be possible to talk in terms of the conservation of energy if, in addition to the particles seen in the cloud chambers, there was an electrically neutral particle that could not be seen but that nevertheless carried off some of the energy. Furthermore, in order for energy to be conserved, this particle would seem to be massless, just as the photon is.[45] The

Italian Nobel laureate Enrico Fermi gave the name *neutrino* to Pauli's hypothetical particle.

Unfortunately, the effects of neutrinos proved almost impossible to observe. It seemed that neutrinos must be able to pass through matter almost as if it were not there. These undetectable particles were a thorn in the side of many physicists. In 1938, Sir Arthur Eddington wrote,

> In an ordinary way I might say that I do not believe in neutrinos. But I have to reflect that a physicist may be an artist, and you never know where you are with artists. My old-fashioned kind of disbelief in neutrinos is scarcely enough. Dare I say that experimental physicists will not have sufficient ingenuity to *make* neutrinos? Whatever I may think, I am not going to be lured into a wager against the skill of experimenters under the impression that it is a wager against the truth of a theory. If they succeed in making neutrinos, perhaps even in developing industrial applications of them, I suppose I shall have to believe—though I may feel they have not been playing quite fair.[46]

Eddington's reluctance to wager and Pauli's faith in the conservation of energy were justified many years later when nuclear reactors were developed. The American physicists Clyde Cowan and Frederick Reines reasoned that if Pauli had been right, the reactors would be emitting huge numbers of neutrinos. Despite the extremely small likelihood that any one neutrino would interact with a proton, the immense number of neutrinos that ought to be flowing from a nuclear reactor meant that a few neutrinos should interact with atoms in the shielding around the core of the reactor. In this process, the theory predicted that high-energy photons would be produced. By finding these photons, Cowan and Reines were able to infer the presence of Pauli's elusive particles and to conclude that neutrinos are as "real" as any other subatomic particle.

New Conservation Laws

So far we have seen two conservation laws: the conservation of energy, illustrated with the roller-coaster and neutrino examples, and the conservation of electric charge. The usefulness of these two notions led physicists to attempt to formulate other conservation laws. One such law is the *conservation of momentum*. For example, before a collision takes place, we can multiply the mass of each particle by its velocity. We can add up these individual contributions and get a number.[47] If we carry out the same calculation after the collision takes place, we will get a second number. The conservation of linear momentum is the declaration that these two numbers will always be the same if there is no outside force acting on the system. Physicists find that *every* process can be described in a way that is consistent with the conservation of momentum.

When Einstein postulated the equivalence of mass and energy, he found that he could continue to talk in terms of the conservation of energy only if he redefined energy to include the energy represented by the mass of a particle ($E = mc^2$). This redefinition also allows physicists to employ the conservation of energy when talking about atomic phenomena, just as they do when talking about the large-scale world. The conservation of momentum can also be applied to the atomic world. It is even possible to continue to talk in terms of the conservation of energy and of momentum when some particles disappear and new ones are created.

Rotational motion can also be described in terms of a conservation law; the principle is then called the *conservation of angular momentum*. Kepler's second law, involving the speed of a planet and its distance from the sun, is now said to be an example of the conservation of angular momentum. Both classical and atomic interactions can be described in this way.

Talk about angular momentum needs to be qualified when the subatomic world is involved, however. If the electron does not have a definite path through space and time, the electron cannot have a well-defined orbit. Then how can we say that it has an angular momentum associated with its orbit? Physicists found that a number measured in the same units as angular momentum could be associated with the "orbit" of an electron. Unlike a planet, which can have any value for its orbital angular momentum, an electron can have only certain values for its orbital angular momentum, which is what Bohr first told us. In addition, this angular momentum can have the value zero, but any planet with an orbital angular momentum of zero would fall into the sun. Nevertheless, talking about this number *as though* it were angular momentum allows physicists to talk about the conservation of angular momentum at the subatomic level in the same way they talk about the angular momentum of merry-go-rounds or planets.

If an astronaut were suspended in space far from the sun or any other star, the way we think of space makes it seem reasonable to say there would be no up or down or any other preferred direction in space. Say the astronaut found an object floating in space and adjusted her speed and direction until she was motionless with respect to the object. We would not expect the object to suddenly start moving in some direction unless a force were applied to it. This expectation is another way of stating the conservation of momentum. Similar arguments can be made for the conservation of angular momentum and the conservation of energy. These conservation laws are closely linked to the way we talk about time and space—we say that space has no preferred direction and that we can perform the same experiments at different times and get the same results. It is not obvious that it *must* always be possible to carry out an experiment and get the same results at different times and places. Fortunately, however, physicists find that the world is somehow put together so that this "symmetry" holds—if it did not, it is hard to imagine that we could ever find laws behind the apparent chaos or that we could even exist.[48]

Physicists looked to see what kinds of interactions occur and what kinds do not occur in the subatomic world and talked about the results of their studies in terms of additional conservation laws. Just as the conservation of energy allowed physicists to rule out forbidden roller coasters, they assumed that whatever they did not see was prevented by other conservation laws.[49] For example, in one form of radioactive decay, a neutron decays into three lighter particles, a proton, an electron, and a particle called an antineutrino. But electrons, protons, and antineutrinos are never seen to decay into other particles. There is no lighter particle that has the same charge as the electron. This electron decay is "prevented" by the conservation of energy and electric charge. But there *are* positively charged particles lighter than the proton.

Physicists could "explain" the failure of the proton to decay by inventing another "charge," similar in many ways to electric charge, that is conserved in the same way electric charge is always found to be conserved. Physicists now speak of protons as the lightest particles in a family of particles they call *baryons*. Baryon "charge" is conserved just as electric charge is conserved, but because there is no force associated with baryon charge, it is normally referred to as *baryon number*. Every particle is assigned a baryon number of either plus 1, minus 1, or zero. Since there is no baryon lighter than the proton, the conservation of energy and baryon number combine to explain physicists' failure to observe proton decay.[50]

The failure to observe other apparently permitted decays led physicists to propose still more conservation laws, although these laws, unlike those of conservation of energy and momentum, could not be applied in every interaction. By the time they were through, physicists had invented more than a dozen conservation laws to describe the outcomes of their experiments.

Inventing the Idea of Symmetry

Symmetry is the concept that we can shift a system in some way and still have it look the same. Squares, for example, display one kind of symmetry; snowflakes another. Circles display an even greater symmetry: No matter how you rotate one, it always looks the same. Yet another symmetry is associated with the observation that if all the positive electric charges in any configuration are replaced by negative charges, and all the negative charges are replaced by positive charges, the energy in the electric field created by the charges remains unchanged. Physicists refer to nature's indifference to how they label electric charge, or where they set the zero point in making measurements, as examples of symmetry. Nature looks the same no matter how physicists set up coordinate systems or what they choose to call the beginning of a circle.

To see how symmetry works, set out three pennies on the table in front of you. Turn one penny so that it is heads up, a second so that it is tails up, and the third so that it is tails up. We can represent this arrangement by HTT. Now turn over the first two coins to produce the arrangement THT. If you now turn over the two end coins, you will have HHH. Notice that if you performed these two actions in reverse order you would have gotten TTH followed by HHH—the same end result you had originally. The order in which you performed the operations did not affect the final arrangement. Furthermore, you could have reached the same final configuration by turning over the second two pennies in the first configuration. The fact that it does not matter in which order you perform the operations and that you can achieve the same final state by a single step or by several intermediate steps are properties defining a set of symmetries—ways of arriving at the same destination despite traveling over different routes.

In 1832, at the age of 20, the French mathematical genius Evariste Galois was killed in duel involving a woman. In his lamentably short career Galois invented a mathematical way of describing the laws that govern symmetry, an area of mathematics now called *group theory*. A group is a set of elements (heads or tails) that can be converted into each other by definite rules (flipping the coin).

In 1918 Amalie Emma Noether, the first woman to become a member of the faculty of the University of Gottingen in Germany, demonstrated a very important way to describe conservation laws in terms of symmetry. Noether showed that symmetries associated with the mathematical expression for the energy of a system will *always* appear in the form of conservation laws. The symmetries in this case ensure that calculations based on the energy will give the same answer no matter what coordinate system physicists use to describe the process or how that system varies from place to place. Noether's theorem provided physicists with a powerful way to understand the conservation laws they had developed. It states that these laws tell physicists what symmetries they must incorporate into the mathematical expressions they write to describe the behavior of physical systems.

The power of the idea of symmetry was not obvious during the 50 years separating the mid twenties from the mid seventies, and as a result the approach developed very slowly during that period. Only from the exalted perspective of hindsight can we see the acceptance of this language as inevitable.

QED Revisited

Greenwich, England, is the zero point from which longitude is measured. If we set up another system in which the zero point of longitude is located in Pocatello, Idaho, the longitude of every city in the world would change, but the *distance* between New York and Boston would remain exactly the same. One way to express this is to say that distance is unchanging, or *invariant,* with respect to changes in the location from which longitude is measured. This invariance is comforting because the distance between the cities is something we could measure in other ways—with a car's odometer, for example—that do not involve longitude and latitude, which are part, after all, of a purely conventional reference system. We believe that any system of geographical reference *ought* to leave the distances between cities unchanged (except for the units in which distance is measured—e.g., miles or kilometers).

The equations of QED have a property similar to the conventionality of geographic coordinates. In the equations describing the energy of electrons, for example, physicists must arbitrarily assign a "phase" to the electron field (as we must arbitrarily decide on the zero point from which longitude is measured). Just as we want the distances between cities to remain the same no matter where we set the zero point for longitude, physicists want the predictions of QED to be independent of the phase they choose. In particular they want electric charge to be conserved. If they confine their equations to terms related to the behavior of electrons and positrons, however, physicists find that their predictions *do* depend on the phase of the particles. To free their predictions from this dependence, they must add a new field to their description. For historical reasons this field is called a *gauge field.* The word *gauge* is sometimes used to describe the distance between the rails of a train line, and we can think of a gauge field as a way of keeping a physical theory on track. There is no term involving mass associated with the gauge particles in QED, and since photons are massless, physicists identify the gauge field in QED with the electromagnetic field. Photons, the "carriers" of the electromagnetic field, are therefore sometimes described as *gauge particles.*

Gauge Fields—An Allegory

Imagine that you are in a town served by a commuter system in which trains arrive every 20 minutes. When the town switches to daylight savings time, everyone sets his or her clocks an hour ahead, but everyone can still predict when the next train is going to arrive. (This is an example of a *global* symmetry—the schedule of the trains remains unchanged if all the clocks are changed by the *same* amount.)

Imagine that the trains are so punctual that each person in the town uses the arrival of a train to set his or her clock. (In this case each person chooses his or her own *phase convention.*) This procedure, however, does not guarantee that any of the clocks show the same time. If one person sets his clock at 3:00 just as a train arrives, and another sets hers at 4:20, each of them can predict when the next train will come (3:20 by the first clock and 4:40 by the second), but neither knows how the other's clock is set. (This is an example of a *local* symmetry—the clocks are all altered by *different* amounts, but the predicted arrival of trains remains unchanged.) In both global and local symmetries, the predicted schedule of the trains does not depend on the setting of any particular clock.

If two townspeople want to rendezvous at the train station, however, they must arrange some way to synchronize their clocks, since there is no reason to believe they are already set to the same time. We can imagine that the townspeople develop a series of signals ("Flash your light when your clock reads noon") to make this synchronization possible. In an analogous manner, the phase of the electron field can be said to be altered by photons, which synchronize the "clocks" of electrons and positrons and thereby permit them to interact.

The interaction between electrons is described in the following way: When two electrons meet, one electron can emit a virtual photon, which is absorbed by the second electron. Physicists describe this process by saying that two electrons interact by *exchanging* a virtual photon. (The way two townspeople can interact is by exchanging a message defining the time—by synchronizing their watches.) This represents yet another way to talk about force: *A force arises when virtual gauge particles are exchanged.*

Using Symmetry

The subatomic world seemed simple in 1927. Physicists talked about two kinds of subatomic particles, the lightweight electron, which carries negative electric charge, and the heavier proton, which carries positive electric charge. They also talked about two forces, gravity and electromagnetism. But from 1927 on, things became more and more complicated. To begin with, there was the problem of explaining how the protons in the nucleus are able to remain bound together despite the fact that their positive electric charges repel each other very strongly.

The nucleus of the atom became more complex in 1932 when, after 12 years of searching, the English physicist John Cockcroft discovered that along with protons the nucleus contains heavy particles carrying no electric charge—neutrons. After repeated failures to describe how the nucleus could be stable if only the electromagnetic force is involved (the gravitational attraction between protons is much too weak to hold the nucleus together), the existence of a neutral particle reinforced the conviction that there must be yet another fundamental force—a nuclear force—in addition to gravity and electromagnetism. This nuclear force must be very powerful in order to overcome the strong electric repulsion of the protons and to bind electrically charged protons to electrically neutral neutrons. Finally, unlike gravity and electromagnetism, the effect of this force must be short-ranged, barely extending for more than 2 or 3 diameters of a proton, or it might pull all nearby protons and neutrons into a giant atomic nucleus. Showing a lack of imagination that later colleagues would make up for, physicists called this the *strong nuclear force*.

In one form of radioactive decay, a neutron disappears, a proton appears, and an atomic nucleus emits an electron. This phenomenon could not be explained on the basis of the known forces, and the Italian physicist Enrico Fermi postulated yet another new force, one that controlled radioactive decay. In Fermi's model an electron is created in much the same way as a photon is created when an atom emits light. His force eventually came to be called the *weak nuclear force*.

Heisenberg pointed out that the force binding the nucleus together does not appear to discriminate between neutrons and protons; nuclear interactions seem to remain unchanged if protons are replaced by neutrons and vice versa. He argued that as far as the strong force is concerned, physicists can talk about protons and neutrons as varieties of the same particle—a particle he called the *nucleon*. We can think of a nucleon as a coin. If the coin is heads up we call it a proton; if the coin is tails up, we call it a neutron. The nucleon, however, unlike a coin, does not rotate in physical space but in an "internal" mathematical space.[51]

Mathematical spaces are abstract analogies to physical spaces. In this case, the analogy arises because the same mathematical rules that can be used to describe the results of rotating an object in physical space can be successfully applied to describe outcomes that have nothing to do with rotations in physical space—in this case with the conversion of a neutron into a proton. For historical reasons, this abstract space is called *isotopic spin*, or *isospin*, space.

As abstract as this approach may seem, we have by now become used to seeing how, with familiarity, abstract ideas become concrete. In the words of the American physicist Anthony Zee, "With the passage of time, Heisenberg's notion of an internal symmetry no longer appears so revolutionary. To later generations of physicists, internal symmetry seems as natural and real as spacetime symmetry."[52]

In 1934 the Japanese physicist Hideki Yukawa attempted to describe the nucleus of the atom using the language of quantum field theory that Dirac had so successfully used to combine quantum mechanics and relativity. In QED the force between electrons and positrons arises as the result of exchanging virtual photons. Yukawa reasoned that it should be possible to talk about the force between nucleons arising from the exchange of some other particle. But what particle? Yukawa knew that Heisenberg's relationship allows physicists to talk of virtual particles with mass only if the particles exist for a very short time, since otherwise such particles would violate the conservation of energy. During this brief interval the particles can travel only a short distance, even if they travel with the speed of light, and since the nuclear force is a short-ranged force, a massive

particle seemed a logical way to describe this force. Yukawa postulated the existence of a new particle that his calculations indicated
should have roughly 200 times the mass of the electron. After several
false alarms, Yukawa's particle (now called the *pion*) was discovered
in 1947. Yukawa's description of the strong force foreshadowed the
way physicists would come to talk about *all* the forces of nature.

The Yang-Mills Language

Chen Ning Yang and Robert Mills, working together in New York's
Brookhaven National Laboratory in 1954, attempted to describe
the strong nuclear force by using the approach that worked so
well in QED—by using the language of quantum field theory and
incorporating the preservation of gauge independence.

Once again, think of the nucleon as a coin, keeping in mind that
the nucleon rotates in an abstract, nonphysical isospin space. If the
coin is heads up, we call it a proton; if the coin is heads down, we call
it a neutron. For example, the nucleus of an ordinary helium atom
consists of two protons and two neutrons, so we can represent the
nucleus by four coins, two with heads up and two with tails up.

What we say about the orientation of a nucleon in isospin space
is arbitrary. (There is no absolute way to tell whether heads or
tails is up because there is no absolute "up," just as there is no
absolutely correct setting for all clocks.) Yang and Mills wanted to
make sure that the orientation in the isospin space that they arbitrarily
assigned the nucleon (analogous to the choice of the phase of an
electron) did not affect the predictions of their theory.[53] (The settings
of the clocks does not affect the train schedule.) They wanted to
do more than this, however. They wanted to allow the convention
identifying protons and neutrons to change from place to place
without affecting their predictions—they wanted isospin symmetry to
be a local gauge symmetry. (The clocks in the village all read different
times, but all can be used by their owners to predict the arrival of the
next train.) An equation that only incorporated nucleons, however,
does not preserve isospin symmetry, just as an equation that only
incorporates expressions for electrons and positrons fails to preserve

charge conservation. To overcome this shortcoming, Yang and Mills had to add terms describing a gauge field that would offset the effects of the arbitrary convention.

If one of the heads-up coins representing the helium nucleus is flipped over (if a proton is arbitrarily relabeled as a neutron), we would have one heads-up coin and three tails-up coins and the original pattern would be altered. The initial configuration can be restored, however, by turning over one of the coins that was originally tails up. The gauge-field "signals" instructing the coins whether to be heads up or tails up are similar to the signals the townspeople use to synchronize their clocks. We can think of the gauge field as a mathematical mechanism to ensure that a coin that flips from heads to tails would emit a particle that would signal a second coin to flip from tails to heads in order to restore the original combination of heads and tails.[54]

Just as the gauge field in QED is interpreted as the electromagnetic force, the Yang-Mills gauge field is interpreted as a force. In the same way that two electrons can be said to interact by exchanging photons, two nucleons can be said to interact by exchanging the new Yang-Mills gauge particles. These gauge particles can be thought of as neutralizing any arbitrary change in the convention distinguishing protons from neutrons by transforming protons into neutrons and vice versa. (In this respect the Yang-Mills gauge particles are unlike the photon, which does not alter the identity of an electron.)

The Yang-Mills language provides yet another way to talk about a force: *a force is a way of preserving local gauge symmetry*. This description hardly sounds like what either Newton or Einstein had in mind when they talked about gravitational force, but Einstein's approach does preserve gauge symmetry. In fact, it was this property that led Hermann Weyl to become one of the first to explore the application of symmetry to physics in 1919. If physicists could make a quantum field model of gravity (they have as yet not been able to), the force of gravity would be said to arise as the result of the exchange of gravitons, analogous to photons in electromagnetic theory.

Since a symmetry of the type envisioned by Yang and Mills involves changing neutral particles into charged particles and vice versa, it turns out that more than one kind of gauge field is needed to bring about the required changes. The particles associated with these fields, unlike the photon, must carry information about electric charge. The mathematics requires a total of three gauge fields, rather than simply one as needed by QED. In the equations of QED, mass is associated with the terms for electrons and positrons, but no mass is associated with the expression for photons, the gauge particles that interact with the electrons and positrons. Yang and Mills hoped that their gauge particles, unlike photons, would turn out to have mass for the same reason Yukawa talked about a massive particle—because the range of the strong nuclear force must be small.

It was difficult to make calculations using the Yang-Mills model because the Feynman diagrams associated with the model are extremely complex. This complexity arises because the Yang-Mills gauge particles, unlike photons, interact with each other. As a result Yang and Mills were uncertain whether the new particles would wind up with any mass when all the calculations were complete.

Eventually it became clear that there are no mass terms for the Yang-Mills particles and that simply writing down a term introducing such a mass destroys the symmetry the equations were designed to maintain. Yang and Mills were free to invent any fields they liked to ensure the symmetry of their equations. But since they were trying to talk about the physical world, they had to interpret these fields as having physical consequences. In this case, to preserve the symmetry of their equations, Yang and Mills seemed to call for the existence of massless particles carrying electric charge.

The immediate and apparently insuperable problem is that massless charged particles do not exist in nature. If they did exist, they would be extremely abundant, just as photons are, and, having electric charge, should be very easy to detect. Thus the prediction of massless charged particles by Yang and Mills posed a real problem. In addition, massless particles would imply that the nuclear force is a long-range force, and physicists knew very well that it is a short-range force. As if this were not bad enough, the Yang-Mills approach might well lead to infinities that cannot be removed and hence render the model useless for calculating anything.

This rather unsatisfying situation described the state of quantum field theory throughout much of the 1960s. At this time there was no agreement among physicists on the best way of talking about the forces of nature, but there was a widespread feeling that looking for symmetry and using the language of quantum field theory, with its predictions of nonexistent massless particles and calculations riddled with infinities was *not* the right approach. In fact, Freeman Dyson, the physicist who showed the mathematical equivalence of Feynman's and Schwinger's way of doing quantum electrodynamics, said, "Those of us who still pursue field theory are gradually becoming an isolated band of specialists. . . . It is easy to imagine that in a few years the concepts of field theory will drop totally out of the vocabulary of day-to-day work in high energy physics."[55] Dyson would prove to be completely wrong, but of course at the time there was no way for him to know this.

10

W, Z Fields

Nature does not appear very simple or unified . . . [but] we can at least make out the shape of symmetries, which though broken, are exact principles governing all phenomena, expressions of the beauty of the world. . . .[56]

STEVEN WEINBERG

Symmetry Disguised

Imagine you are balancing a pencil on its point on a tabletop. There is no obvious reason that the pencil should fall in any particular direction because there is no preferred direction on the table as far as the pencil is concerned. But sooner, rather than later, the pencil will topple and wind up pointing in *some* direction. Physicists describe this behavior as *spontaneous symmetry breaking*. When the pencil is lying on the table, the original symmetry of the situation from which the pencil started is hidden. Spontaneous symmetry breaking provided physicists with a way to circumvent the problem of massless particles that cropped up whenever they talked about forces in the language of Yang and Mills.

In the 1960s and early 1970s, the American physicist Steven Weinberg and Pakistani physicist Abdus Salam independently worked out ways to use the language of quantum field theory to talk about a combination of the weak force (the force responsible for radioactive decay) and the electromagnetic force referred to as the *electroweak force*. It turned out to be easier to build a quantum field model of the weak force by including the electromagnetic force as well. Weinberg and Salam took essentially the same approach as Yang and Mills did, calling on a single coin to represent an electron when it is heads up and a neutrino when it is tails down.[57] The coin once again rotates in an abstract mathematical space exactly analogous to the space in which the nucleon can be said to rotate and reveal itself as a proton

137

or a neutron.[58] Like Yang and Mills, Weinberg and Salam needed a way to ensure the symmetry of their expression for energy, and this implied the existence of three new gauge particles. The equations had no terms incorporating the mass of the new particles; like the Yang-Mills particles and the photon, the new particles were massless.

At this point Weinberg and Salam were no better off than Yang and Mills, since all four predicted the existence of massless charged particles that should be easy to observe but are never seen. Weinberg and Salam, however, found that by incorporating additional fields into their equations they could add mass to the otherwise massless particles. The fields are named *Higg's fields* after their inventor, the Englishman Peter Higgs, and can be thought of as a device for concealing the massless nature of the gauge particles in the electroweak theory. The price for adding these new fields is the creation of a new class of particles called *Higgs particles*.

Weinberg and Salam's mathematics implies that the Higgs fields interact with the massless particles needed to ensure the symmetry of the equations. In the process, these massless particles gain mass, and the massive Higgs particles that would otherwise be predicted by the theory vanish. As Salam describes it, the gauge particles eat the Higgs particles to gain weight, and the Higgs particles become ghosts.

The Higgs mechanism works something like the wise man's camel in an old story. According to legend, a desert chieftain left 17 camels to be divided among his three sons. His instructions were that his oldest son was to inherit half of his camels, his middle son one-third, and his youngest son one-ninth. No matter how they tried, the elders of the tribe could not fulfill the chieftan's wishes without cutting up one or more of the camels—a clearly undesirable outcome. Then a wise man came to the village and tied down his camel next to the chief's camels. Now there were 18 camels. The wise man gave the oldest son half of these (nine camels), the middle son one-third (six camels), and the youngest son one-ninth (two camels), and lo, one camel was left. The wise man mounted the remaining camel, which of course was his, and rode off. Like the wise man's camel, the Higgs particle is needed to make Weinberg and Salam's mathematics work out. At the end of the calculation, however, the particle, like the extra camel, often disappears.

Combining Forces

Several years before Weinberg and Salam did their work, the American physicist Sheldon Glashow had used a similar approach to talk about a combination of electromagnetism and the weak force. To avoid the problem of massless charged particles, Glashow had added mass to the particles in his equations by hand; that is, he introduced terms for the mass without having something like the Higgs mechanism to account for the origin of the mass. In doing this, Glashow destroyed one of the results he was trying to preserve—the underlying symmetry of the model. Weinberg and Salam hoped that the Higgs mechanism would not destroy the symmetry but only hide it, because the massless particles the symmetry leads to (like the Yang and Mills photons with electric charge) are never seen. They also hoped that the Higgs mechanism would lead to models in which the infinities that plague quantum field theory could be defined away just as they are in QED.

The Weinberg-Salam approach calls for the existence of three new particles. Two of these particles, the W^+ and the W^-, carry electric charge. The third, the Z^0, carries none. These three particles start out massless, but they gain mass by the Higgs mechanism. In fact, they gain quite a bit of mass—winding up almost 100 times as massive as a proton. The Z^0 can be described as a *very* heavy photon, because, like the photon, the Z^0 has no electric charge. In the mathematical mechanism by which these particles gain mass, three varieties of Higgs particles vanish. But the model requires a fourth Higgs field, and since the photon has no mass, it does not cancel out the mass of the fourth Higgs particle, which therefore remains at the end of the calculations. By adding the Higgs fields to their models, Weinberg and Salam were predicting the existence of *four* new particles—the two W's, the Z, and the Higgs.

Physicists produce particles by smashing other particles into each other. In accordance with Einstein's equivalence of mass and energy, the heavier the particles produced, the more energy it takes to move them. But the highest energy accelerators available in the late 1960s had insufficient energy to produce the putative massive W's and Z and the equations did not allow Weinberg or Salam to predict what the mass of the Higgs particle might be.

In science, a good measure of the impact of a paper is given by the number of times it is referred to in publications by other authors in the same field of research. In the four years following its publication, Weinberg's paper laying out the electroweak unification was cited only once. Hardly an auspicious welcome for the first step since Maxwell's in unifying the forces of nature in one theory. Why such indifference?

First, the massive W^+, W^-, Z^0, and Higgs bosons had never been observed, nor could they be made with existing accelerators. Second, the existence of the Z^0 particle implies that there must be weak interactions in which electric charge does not change, just as the neutral photon allows electrons to interact without changing their electric charge. But, despite extensive searches, no such "neutral current" weak processes had ever been observed. Third, although the neutron and the proton are very similar in mass, and hence reasonable candidates for being related, the massless photon and the very massive Z^0 hardly seemed to be fundamentally identical. Finally, there was no reason to believe that Weinberg and Salam's model avoided the problem of infinities and hence would be useful in making calculations. Weinberg and Salam hoped the infinities could be absorbed by some redefinition, but they offered few reasons for their hope.

A variety of approaches to describing the strong and weak forces enjoyed some popularity, but no dramatic advances were made. Many physicists were depressed by the repeated failures to make any progress in finding a fundamental way to talk about nature. Even Weinberg turned his attention to other questions.

By the end of the 1960s, most physicists had lost interest in the language of quantum field theory. Its problems seemed insuperable. One exception, however, was the Dutch physicist Martinus Veltman, who found that some of his field theory calculations seemed to call for the existence of particles with negative probabilities of existence. As bad as negative energies were for Dirac, negative probabilities seemed worse (what could it mean to have *less* than *no* chance of appearing?). Veltman found that these particles produce no problems, however, as long as they vanish *before* the particles could be expected to show up in experimenters' laboratories. The particles, in other words, must always remain virtual. These "ghost" particles gave physicists a way to avoid being committed to the "existence" of some particles that their mathematics otherwise seems to require.

In 1971 Veltman was working with Gerard 't Hooft, then a graduate student at the University of Utrecht. Much to Veltman's surprise, 't Hooft succeeded in demonstrating several important points that had thwarted other physicists. First, he showed that the infinities arising whenever physicists use the Yang-Mills language, like those in QED, *can* be redefined away. Equally important, 't Hooft showed that this redefinition is still possible when the Higgs mechanism is used to give mass to the gauge particles required by the Yang-Mills language. Weinberg and Salam's hopes were realized: Their models *might* be as useful as QED.

Following the report of 't Hooft's work, few were willing to accept his conclusions—after all, 't Hooft, a mere graduate student was claiming to have solved a problem that had stymied some of the best minds in physics for decades. But when the work of the well-known Korean-born American physicist, Benjamin Lee, confirmed the findings, theorists began to take 't Hooft's conclusion seriously. The language of quantum field models again became fashionable, and soon the models glutted the market. But which, if any, were useful in talking about nature? Only experiments could answer this question, and experimental physicists began to search for the types of atomic interactions called for by the theorists.

By 1976 a variety of experiments had revealed phenomena consistent with the Weinberg-Salam theory, including the long-sought neutral currents. Although there was still no direct evidence for the existence of the W and Z particles, Glashow, Salam, and Weinberg were awarded the 1979 Nobel Prize for their work on the theory of the electroweak force.

There was still the problem of the W and Z particles. The W particle would decay much too rapidly to be identified in any detector.

Instead, physicists would have to infer the existence of the W by identifying the particles the W produces when it decays. In 1983 the Italian physicist Carlo Rubbia, with the support of dozens of scientists and engineers, successfully used a new accelerator at the European Center for Nuclear Studies (CERN) to create particles with inferred masses equal to those predicted for the W and Z particles by the electroweak theory.[59] The Higgs particle has still not been seen, but since the model gives few clues to what its mass might be, its possible that the particle simply is too massive to create in existing accelerators.

Just as the development of QED required physicists to invent the idea of empty space teeming with unobservable particles, the electroweak theory required the invention of the unobserved Higgs fields and their associated particles. If evidence could be found for the existence of Higgs particles, some physicists would be more comfortable with this approach, because at present the only role of the Higgs field is to make the mathematics of the electroweak theory work out. However, the Higgs particle is only one of a number of unobserved particles. Many of these fundamental particles, although they are not ghost particles, are unobservable even in principle. How did physicists get themselves into such a predicament?

11

The Ineffable Color of Quarks

Science is the attempt to make the chaotic diversity of our sense-experience correspond to a logically uniform system of thought.[60]
ALBERT EINSTEIN

Someone once likened studying the nature of matter by using a particle accelerator to studying the mechanism of a watch by smashing it against the wall and looking to see what pieces fly out. Actually, the situation would not be too bad if it were really that good. Presumably, when we smash a watch against the wall, the pieces that fly out were in the watch all along. In a particle accelerator, the pieces that fly out were *not* there to begin with: Most are created out of the energy involved in the interaction. In the language of field theory, they are particles that acquire enough energy to be promoted from virtual to real status. In this sense, physicists *create* new particles rather than simply discover them.

A *particle accelerator* is a device that allows physicists to transfer a great deal of energy to subatomic particles. In some designs, streams of either protons or electrons are directed against targets, and the resulting shower of particles is then studied with a variety of detectors. In the most powerful accelerators, beams of electrons collide with beams of positrons; in still others, beams of protons collide with their corresponding antiparticle, antiprotons. In these collisions the mass of both particles is converted to energy in accordance with Einstein's relationship between mass and energy. Out of this energy, new particles are created.

One problem with using particle accelerators to study the makeup of matter is that the same pieces do not always fly out—some particles are more likely to emerge than others. As time has gone by, physicists have become increasingly interested in rare events, and these events can require months and even years of gathering data to uncover. The result of smashing two particles together can be observed in a bubble chamber, where the resultant particles reveal themselves by the tiny trails of bubbles produced by their passage through the chamber, much as tracks are left in cloud chambers.[61] Many particles are not seen directly, either because their lifetimes are very short or because they are not electrically charged and do not leave a path in the bubble chamber.

During the process of carrying out an experiment, tens of thousands of bubble-chamber photographs may be produced, each one of which must be carefully examined for the patterns characteristic of the particle being sought. The situation is further complicated by the fact that different particles may produce similar patterns, and statistical methods must be used to determine the most likely cause of a pattern. The studies we will be looking at can easily involve hundreds of people and take years to complete. With this brief introduction, let us now look at how physicists found a way to talk about the force that holds the nucleus of the atom together.

The development of particle accelerators, working at ever higher energies, allowed physicists to create still more new particles—eventually some 200 of them. The world of fundamental particles has become very complicated indeed. To make some sense out of this profusion, physicists turned to a familiar and useful approach—conservation laws.

The Eightfold Way

Although physicists had invented a number of new conservation laws, they were still far from having a coherent way of talking about the hundreds of subatomic particles. Then, in 1961, Murray Gell-Mann and the Israeli physicist Yuval Ne'eman independently proposed a scheme that brought some order to the profusion of particles. They found a way of introducing symmetry into the confusion by arranging many of the particles then known into orderly families based on the conserved quantities found to be associated with each particle. Since the first family in this structure had eight members, Gell-Mann flippantly called the system *the eightfold way*, after Buddha's principles for right living.

The eightfold way not only allowed existing particles to be characterized; it allowed Gell-Mann and Ne'eman to predict the existence and mass of a new particle, the Omega-minus. In 1964, after a search requiring the examination of over a million feet of bubble-chamber photographs, evidence for a particle with the characteristics of the Omega-minus was found. Physicists were now in a situation much like the one that existed after Mendeleev proposed the periodic table of elements in 1869. They had a scheme that resembled the table but no idea of what enabled this recipe to work.

The Coming of Quarks

In 1964 Gell-Mann and, working independently, the American physicist George Zweig saw a way to explain the symmetries involved in the eightfold way. They decided that the fundamental symmetry had to involve particles that had never been observed. Gell-Mann called these particles *quarks*—a name he took from James Joyce's *Finnegans Wake*, because he liked the sound of it—and Zweig called them *aces*. Quarks or aces, however, the particles had the unfortunate characteristic of carrying fractional electrical charge. Gell-Mann, at least, was reluctant to draw too much attention to this conclusion.

It is difficult for nonphysicists to imagine just how little physicists liked the idea of anything with less than the charge of a single electron. The idea of quarks, in Gell-Mann's terminology, or aces, in Zweig's, was *so* unpopular that Zweig recalled:

> The reaction of the theoretical physics community to the ace model was generally not benign. Getting the . . . report published in the form that I wanted was so difficult that I finally gave up trying. When the physics department of a leading university was considering an appointment for me, their senior theorist, one of the most respected spokesman for all of theoretical physics, blocked the appointment at a faculty meeting by passionately arguing that the ace model was the work of a "charlatan".[62]

Why such antipathy? The electric charge of an electron is exactly equal to and opposite in sign to the charge of a proton. Whenever electric charge is found in nature, it is always found in exact multiples of the charge of the electron or the proton. Gell-Mann and Zweig suggested that the fundamental constituents of matter carry charges of only one-third and two-thirds the charge of the electron. Many precise measurements had been carried out of the charges of subatomic particles, and they had *never* revealed any particle with a charge less than the charge on the electron.[63] Since particles carrying fractional electrical charge have never been observed, most physicists paid little attention to the quark approach, which to many seemed to be little more than a clever mathematical scheme, certainly not a believable physical description. Still, in physics what is unbelievable today has a way of becoming compellingly obvious tomorrow—and this was to be the fate of quarks.

If the problem of fractional charge were not enough, the quark language suffered from other difficulties. In Gell-Mann's model, three different types of quarks were needed to describe all the then known varieties of subatomic particles. Furthermore, some particles, such as the Omega-minus, are made up of three of the same type of quarks. Each quark can be thought of as spinning like a top with the same amount of spin. The three quarks that make up the Omega-minus particle are confined to the same location, and to account for the spin of the Omega-minus, all three quarks must have their spins pointing in the same direction. Thus these quarks seem to violate a principle first articulated by Wolfgang Pauli that says, among other things, that no two identical particles making up matter can occupy the same place at the same time if they also have the same spin direction.

The Japanese physicist Yoichiro Nambu found a way around the similar quark difficulty; he suggested that the three quarks are not quite identical. If we go back to our analogy and think of quarks as coins, Nambu suggested that each coin could also have a different color. For example, the Omega-minus particle would consist of three similar quarks, one "red," one "green," and one "blue." The colors of the three quarks are not identical, and the Pauli principle is therefore not violated. "Color," of course, has nothing to do with color; it is simply a way of labeling a mathematical relationship.

Nevertheless, their antipathy to quarks led physicists to look to other vocabularies and other languages. Some talked about the strong interaction in terms of mathematical expressions, called *Regge poles,* that linked particle masses and spins. The American physicist Geoffrey Chew argued in the 1960s that no physical particle was more fundamental than any other—each particle helps generate other particles, which in turn generate it. This *bootstrap hypothesis* enjoyed some popularity as an alternative to the language of quantum field theory. The bootstrap language was an attempt to break away from the way physicists had successfully talked since the time of Galileo: describing phenomena on one scale in terms of phenomena on an underlying scale (e.g., describing the behavior of a gas in terms of the behavior of atoms). This apostasy, however, was to be short-lived.

Just as 't Hooft's demonstration that infinities can be dealt with in models incorporating the Higgs mechanism gave new life to the Glashow-Weinberg-Salam model of the weak interactions, 't Hooft's work gave new life to a quark model formulated in the same language. Gell-Mann realized that the addition of color to his model would solve the Pauli problem and allow him to describe a quark model with the still not very popular language of quantum field theory.

In 1972 Gell-Mann proposed a way of talking about the strong force based on the existence of three kinds of quarks. These quarks can be thought of as corresponding to heads and tails of the nucleon (since there are three quarks, the coin must have three sides).[64] Gell-Mann gave the name *up* to heads, *down* to tails, and *strange* to the third face. He called these the *flavors* of quarks. (It is hard to imagine that a company would flourish trying to sell up, down, and strange ice cream.) The rotation that transforms one quark into another (heads into tails into other) takes place in an abstract mathematical space. Quarks also come in three "colors": "red," "green," and "blue." We can imagine that each coin, in addition to being face up, face down, or other side up, can be oriented so that, if the coin is a penny, the quark is "red" when Lincoln's head points to 12 o'clock, "green" at 4:00, and "blue" at 8:00. This rotation must be imagined to take place in an abstract "color" space.

The metaphor of color provided a clever way for Gell-Mann to explain why physicists do not find objects composed of four or more quarks. Gell-Mann said that nature is arranged so that only colorless particles are stable. Some particles consist of a red, a green, and a blue quark and hence are white, or colorless (this is how a color television set produces white). There are also anticolors, so that some particles consist of one quark of a color and one of the corresponding anticolor—say a red quark and an anti-red, or cyan, quark—and this combination is also said to be colorless. By speaking in this way Gell-Mann argued that physicists never see color or its effects, just as he had argued that nature must somehow be arranged so physicists can never see fractionally charged quarks.

Yang and Mills, as well as Weinberg and Salam, built their formulas on the preservation of a symmetry. Since Gell-Mann wanted to take the same approach, he needed a way to preserve symmetry when the color assigned to individual quarks changes. Since color is a convention, it is arbitrary which color is assigned to which quark. The force between quarks can be thought of as a mechanism for balancing colors. This force arises, as do electromagnetism and the weak force, by the exchange of particles.

Imagine that one of the quarks making up a proton is rotated in color space and converted from red to green. Such a transformation would leave the proton with two green quarks and one blue quark—a combination that is neither colorless nor identical with the original color combination. The new combination would remain colorless and symmetry would be restored if the first quark emits a particle that informs the quark that was originally green to rotate into the red position. If this happens, the proton would again be colorless. This "messenger," or gauge, particle plays exactly the same role as the photon in QED.

The color force is blind to the flavor of quarks just as the strong force is blind to whether a nucleon is a proton or a neutron. Gell-Mann called the particles that carry the color "charge" *gluons,* because in the language of quantum field theory, quarks are glued together by exchanging these particles.

Gell-Mann's mathematics requires eight varieties of gluons to account for all possible color changes.[65] The color force between quarks arises from the exchange of gluons, just as the electromagnetic force between electrons arises from the exchange of photons, and the weak force arises from the exchange of W and Z particles. Since color is the central "force" in the theory, Gell-Mann called his theory *quantum chromodynamics,* or *QCD.*

Still, quarks remained something of a theoretical curiosity. In the process of creating a way to talk about the symmetry underlying the eightfold way, Gell-Mann had been forced to invent particles that had never been observed. It was particularly difficult to explain why, if the theory was to be taken seriously, fractional electric charge and massless gluons are never seen.

At first physicists paid little attention to Gell-Mann's model, but then a series of experiments at the newly built Stanford Linear Accelerator Center (SLAC) revealed unexpected properties of protons. At low energies, electrons scatter from protons in much the same way alpha particles did in Rutherford's famous experiment that revealed the atom has a nucleus; at SLAC the nucleus was simply a proton. At higher energies, however, a new phenomenon began to appear. This new behavior made sense if the electrons were scattering from small pointlike particles *inside* the proton. Feynman called these particles *partons,* but Gell-Mann was sure they were quarks. A major problem with the partons/quarks, however, was that they appeared to wander about quite freely inside of protons. If they were quarks, such independent particles should seem to be easily knocked out of the proton. And if quarks could be easily knocked out of protons, why couldn't physicists see evidence for fractional electric charge?

The few physicists who were enthusiastic about quarks were unwilling to give them up easily. These physicists felt there must be something unusual about the force that holds quarks together. To explain the facts that quarks inside a proton are free to move about and that free quarks are never seen, physicists argued that the force binding quarks together must somehow grow *stronger* the further apart the quarks are. There was one problem with this solution—no one had ever seen a force that behaves in this way.

In 1972 an American physicist, David Gross, attempted to put an end to talk about a quantum field theory of the strong force once and for all by proving that *no* models that include forces that grow stronger with distance can be constructed by using the language of quantum field theory. But a funny thing happened to him on the way to his proof. Gross, 't Hooft, and the Americans Frank Wilczek and David Politzer all succeeded in demonstrating that there is *one* class of quantum field theories that do embody forces that grow stronger with distance—models built in our old friend the Yang-Mills language. In Gross's words, "It was like you're sure there's no God, and you prove every way that there's no God, and as the last proof, you go up on the mountain—and there He appears in front of you."[66] There was no longer any doubt that quantum field theory was an extremely powerful language for talking about nature.

Imagine that quarks are tied together by an elastic string. As long as the quarks stay close together (inside a region we call a proton, for example), the string allows them to wander about freely. When two quarks try to separate a greater distance, however, the string binds them firmly. The same mechanism presumably keeps physicists from seeing the massless gluons, which are safely tucked away inside the particles we can see.

We will return to the question of the nature of reality in a Chapters 14 and 15, but for now the growing acceptance of quarks provides a good example of how a physicist's picture of reality is determined. A way of talking about nature becomes more than *just* a model for a physicist when it becomes apparent that the language has a range of applicability that extends far beyond the problem the language was invented to solve. The most unlikely ways of talking about nature are adopted when physicists find that the newer language helps them solve problems.

A Language Vindicated

Heisenberg told us that in the same way that there is a relationship between position and momentum, there is a relationship between energy and time. We can know one very well only at the expense of knowing the other less well. You may recall that this relationship allows the "existence" of virtual particles, which can "borrow" energy if they pay it back rapidly enough. Heisenberg's relationship entails that we must take time to measure energy accurately. "New" particles are discovered in particle accelerators in the form of a burst of "old" particle production when the accelerator is running at some particular energy. These abundant particles presumably result from the decay of the new particle, and they normally have a broad range of energies. Heisenberg's relationship tells us that this broad energy range reflects the extremely short lifetimes of most subatomic particles—lifetimes usually much too short to be observed directly.

In 1974, Samuel Ting at Brookhaven and Burton Richter at Stanford announced the simultaneous discovery of a subatomic particle characterized by an extremely narrow energy range and thus, by Heisenberg's relationship, an extremely long (by subatomic particle standards) lifetime. The long lifetime implied that the particle cannot decay by the most common mechanisms, since these lead to rapid decay. Several years before, Glashow, Iliopoulos and Maiani had proposed a way to resolve a fairly obscure problem in describing the weak force: They suggested that there must be yet another flavor of quark. Glashow, evidently competing with Gell-Mann to see who was the more whimsical, named it *charm*. Charm seemed like more than just a clever idea for solving an isolated problem, however, when it provided a mechanism to explain the behavior of the new particle. The J/psi, as it came to be called in order to avoid offending either of its creators, was most easily described as a charmed quark bound to a charmed antiquark. Soon evidence emerged for the existence of related particles representing more energetic states of the charmed quark-antiquark pair. These discoveries, sometimes referred to as the "November Revolution," swept away the remaining resistance to quarks. The new vocabulary was now firmly entrenched.

Physicists came to embrace the idea of quarks as much as they embraced the ideas of atoms and virtual particles, and the reason for their conversion was to be the same—the vocabulary is just too fruitful to abandon on the grounds that the particles involved cannot be observed. The more physicists talked about quarks, the more the utility of the notion became apparent and the more comfortable they became with accepting these inferred constituents of the universe. A mathematical abstraction became a physical reality. Physicists had traveled the long road, in Glashow's words, "from mere whimsy to established dogma."[67]

Physicists can now talk consistently about forces in terms of the preservation of gauge symmetry, but they have had to choose carefully the systems that preserve this symmetry. In QED, the symmetry can be demonstrated without much difficulty and involves the existence of the photon. To build the electroweak theory, Weinberg and Salam had to find a way to hide the symmetries in their equations that lead to particles that are not seen in nature, and they invoked a mathematical device called the Higgs mechanism. In the process, two fundamentally identical particles became the massless photon and the supermassive Z particle. Evidence for the W and Z particles has been produced in high-energy accelerators, but as yet there is no evidence for the Higgs particle. To apply the language of gauge symmetry to the strong force, Gell-Mann had to argue that the symmetry exists at a totally unobservable level of reality—the realm of quarks, gluons, and color forces.

Physicists say that quarks never will be observed. They are differentiated by their six flavors (top and bottom won out over truth and beauty as the names for the next two varieties of quarks, demonstrating that physicists are not entirely frivolous); the flavor of the quarks making up a particle determine what kind of particle it is. Only up and down quarks are needed to describe matter on the everyday scale; the other flavors of quarks make their presence felt only in particle accelerators.

Quarks come in three colors. (*Color* is a bookkeeping device that describes the relationships among quarks. Neither flavor nor color can be observed directly; both are conventions.) The weak force is said to change the flavor of quarks by rotating them in *flavor space;* the color force changes the color of quarks by rotating them in *color space.* The color force is carried by gluons, whose function is to make sure that color is never seen. Physicists now talk about the force that holds protons and neutrons together, the strong force, as a residue of color interactions taking place inside protons and neutrons.

Physicists have created a language in which the most fundamental constituents and symmetries of the world cannot be observed. Instead they are properties of operations in abstract internal spaces. Poor Mach, who was bothered by the fact that atoms could not be observed, must be spinning in his grave—and in the process doubtless defining a new direction in yet another abstract space.

I have been talking about the subatomic world in (almost) everyday language, but it is important to remember that the language of physics is mathematics. Only by keeping this in mind is it possible to realize that in the following quotes, two contemporary Nobel laureates are talking about the *same* language (and the same universe!):

I want to emphasize that light comes in this form—particles. It is very important to know that light behaves like particles, especially for those of you who have gone to school, where you were probably told something about light behaving like waves. I'm telling you the way it *does* behave—like particles.[68]   Richard Feynman

Thus the inhabitants of the universe [are] conceived to be a set of fields—an electron field, a proton field, an electromagnetic field— and particles [are] reduced to mere epiphenomena. . . . This picture represents a nearly complete triumph of the field over the particle view of matter. . . ."[69]   Steven Weinberg

The language of quantum field theory is now firmly in place, but with its acceptance the distinction between particles and fields has all but vanished. In Dirac's words:

In atomic theory we have fields and we have particles. The fields and particles are not two different things. They are two different ways of describing the same thing—two different points of view. We use one or the other according to convenience.[70]

Particles and fields are different ways of talking about the world, not different subatomic tinker toys out of which the world is built. The particles physicists now talk about hardly resemble Democritus's atoms or Thomson's electrons. Weinberg went so far as to say, "The particle is nothing else but the representation of its symmetry group. The universe is an enormous direct product of representations of symmetry groups."[71] If anyone needed a demonstration of the power of language, here it is; Weinberg, like the rest of us, identifies the universe with his way of talking about it!

As Lewis Carroll foresaw, reality is becoming curiouser and curiouser. The important point, once again, is that the wildly unintuitive ways that physicists talk are accepted because they confer the ability to organize what otherwise apparently makes no sense. They are accepted because, by and large, they are useful tools. But as 't Hooft warns, "Even where the gauge theories are right, they are not always useful."[72]

The mathematics of QCD is so complex that it is difficult to calculate any number to an accuracy of better than one part in ten, compared to the one part in a hundred million that QED calculations permit. QED computations involve small corrections to an initial approximation. QCD calculations unfortunately involve large corrections—much too large to be treated as simple adjustments to an answer that is close to being correct to begin with.

In Feynman's words, "Here we have a definite theory and hundreds of experiments, but we can't compare them! It's a situation that has never before existed in the history of physics."[73] Physicists hope that this situation is temporary, but for the time being there does not appear to be a way out of this dilemma. Still, this hurdle has hardly slowed down physics at all.

12

The Unquestionable Imagination of Physicists

Our existing theories work well, which is certainly a reason to be happy; but we should also be sad because the fact that they work so well is now revealed as very little assurance that any future theory will look at all like them.[74]

STEVEN WEINBERG

With a common language describing the electromagnetic, weak, and color forces, it was natural that physicists would seek to link all three forces in the same way that the electroweak theory links the first two. There was no reason to seek this unification except aesthetics; physicists would feel better if three apparently disparate forces could be seen as essentially one force. Theories that are built in the attempt to bring about this unification are called *grand unified theories,* or *GUTs.* In 1974, Sheldon Glashow and Howard Georgi developed the first GUT model based on a mathematical structure that encompasses both QCD and the electroweak theory.

If the three forces can be described as manifestations of one hidden force, that "superforce" would arise from the exchange of particles playing a similar role to that played by photons and gluons. Just as the gluon can "transform" a quark of one color into a quark of another color, the superforce particle would be capable of transforming a quark into a lepton. (Matter comes in two varieties, quarks, making up protons and neutrons, and leptons, including electrons and neutrinos.) In the electroweak theory the otherwise massless symmetry-maintaining particles (the W and Z particles) are given mass by Higgs fields. In GUTs the massless particles that maintain symmetry between the color and electroweak forces are also given mass by Higgs fields. Since physicists never see the results of a quark changing into a lepton, they have concluded that such transformations must be very rare. This means that the superforce must be very weak and its range must be very short. The particles associated with the superforce must therefore be very massive—too massive to create in any foreseeable atomic accelerator. The energy at which these particles would be abundant is called the *GUTs energy* or *unification energy*.

The existence of the X particle would imply that quarks can change into leptons. But if the quark that is transformed into an electron or a neutrino is part of a proton, the proton would decay into a lepton and another particle consisting of the two remaining quarks. Thus GUTs predict that protons are unstable and that, rather than enduring forever, matter must eventually decay. If proton decays really occur, their effects would be easy to observe, so they must be exceedingly rare. In order to see why proton decays must be rare, it is necessary to understand what the lifetime of a radioactive particle means.

The lifetime of a particular kind of unstable particle is measured in terms of the time it takes the particles in a sample to decay. Which particular particle will decay at any time is completely unpredictable, just as who will die in a highway accident next year is unpredictable. The lifetime of particles produced in particle accelerators can be as short as a billionth of a trillionth of a second. The lifetime of radioactive cobalt is 77 days. The lifetime of uranium is 4.5 billion years—roughly the age of the earth. If the lifetime of the proton were "only" a million times longer than the age of the universe (roughly 10 billion years), we would be killed by the radioactivity produced by the decay of protons in our bodies! So the life of an average proton must be much longer than a mere 10 million billion years. How could anyone even imagine measuring a process that takes this long?

The answer lies in statistics. When you are 90, your chances of dying in the next year are much greater than they are when you are 30. Radioactive particles are not like that; physicists believe that each radioactive atom has exactly the same probability of decaying in any year. If the time it takes for half of a sample of a radioactive element to decay is a million years, some of the atoms will decay much sooner. How many? A few each year if we have a 10 million atoms. In order to look for proton decay physicists have to put a lot of protons together and watch carefully to see if any decay. One thousand tons of water contains roughly 50 million trillion trillion protons, and if the simplest GUT is correct, roughly five of those should decay in any given year. In other GUTs the rate of decay is even lower. The measurement is, not surprisingly, very difficult to make and the material must be located far underground so cosmic rays cannot penetrate the tank and produce signals that mimic proton decay.

The search has been going on for several years now in several countries. So far there is no evidence for proton decay. The failure to find evidence for such decay has ruled out the simplest GUT. Still, the search continues, and will likely continue for many years, since proton decay represents one of the few tests of GUTs that physicists can hope to carry out.

Even if the electroweak and color forces could be combined in one description, there would still be the challenge of incorporating gravity. Although no discrepancies have yet been found between observations and the predictions of Einstein's theory of gravity, most physicists believe the theory must ultimately break down on scales 100 billion billion times smaller than the proton. General relativity is not a quantum field theory, and on very small scales the quantum effects associated with gravity cannot be ignored if physicists want to be able to apply the language of quantum field theory. However, no one has succeeded in developing a quantum theory of gravity, since no one has been able to avoid the insuperable obstacle of infinities that cannot be removed mathematically. Furthermore, the energy at which the general theory of relativity seems likely to break down is higher even than the GUTs energy. Nevertheless, attempts have been made to unify gravity and the other forces of nature. One such attempt is called *supersymmetry*.

Supersymmetry

If you think we have encountered a lot of fundamental particles so far, you will not be comforted to learn that supersymmetry proposes that every particle in the standard model has an unobserved partner. Corresponding to the photon there is the *photino;* likewise there are *gluinos, winos, zinos, Higgsinos,* and *gravitinos.* There are also *squarks* and *sleptons* corresponding to quarks and leptons. Since none of these supersymmetric partners has ever been observed, why would physicists pay any attention to such a baroque elaboration of an already

161

profuse number of "elementary" constituents of the universe? Super-symmetry promises to provide a theory linking the two fundamental types of particles, the particles: making up the bulk of matter (leptons and quarks) and those that make up the forces (photons, W's and Z's, and gluons). Supersymmetry also holds out the hope of combining gravity into the same model with the other forces of nature. This promise, however, involves the need to talk about ten dimensions of space and one of time. Seven of these spatial dimensions must be hidden somewhere, since we certainly do not see them. Physicists get around this sticky point by arguing that the extra dimensions are "curled up" into spaces much smaller than the size of the proton and so safely invisible.

Physicists explain the absence of evidence for supersymmetric particles by arguing that the particles must be very massive. Once again Higgs fields are invoked to explain this mass. Furthermore, the super-symmetric particles would only weakly interact with normal matter, and so evidence for these particles, like evidence for neutrinos, would be the apparent nonconservation of energy. The possible existence of massive supersymmetric particles is one reason physicists argue that larger atomic accelerators are needed—to attempt to create these particles.

Superstrings

In 1970 Yoichiro Nambu pointed out that one of the then popular ways of talking about the strong interaction is equivalent mathematically to talking about the interactions between one-dimensional objects, or pieces of string. Nambu's model was plagued by the now familiar difficulty facing models of the strong force—it called for the existence of massless particles. In 1974 the French physicist Joel Scherk showed that this difficulty of string theory could be turned into a virtue. Scherk and the American physicist John Schwarz identified the massless particles in string theory with particles that *are* massless—photons and, presumably, gravitons. Scherk and Schwarz argued that string theory, rather than being a theory of the strong force, was properly a theory of *all* the forces—a theory of everything. This change in viewpoint required that the strings be much much smaller than protons, the size that Nambu had in mind. Superstring theories avoid the problem of infinities that have plagued physicists since the early days of QED because strings are not point sources, and although they are very small, they are finite in size.

In their enthusiasm over the successes of the quark language, most physicists paid almost no attention to the string language. Nevertheless, Schwarz continued to explore the power of the new vocabulary. In 1984 he and Michael Green showed that a certain class of superstring theories is free from problems that would have rendered the theories inconsistent with some of the conservation laws, and physicists began to become interested in the new vocabulary. In fact, they became so interested that in the last few years superstring theory has become one of the most fashionable areas of theoretical elementary particle physics.

If string theory is to be consistent, the one-dimensional strings cannot reside in our familiar world of three dimensions of space and one dimension of time. Instead they require nine dimensions of space and one of time. Presumably these extra dimensions are curled up in the way that they are in supersymmetric models. If you find this concept difficult to accept, or even understand, you are not alone. Still, physicists have adopted even stranger notions. As we have seen, extraordinary concepts can win eventual acceptance *if* they solve otherwise intractable problems.

However, there appears to be no way to compare superstring models with observations—all of the distinguishing characteristics of superstrings seem to emerge only at energies much higher than physicists can ever hope to attain in any foreseeable accelerator. Once again, the language physicists find the most powerful seems to hide the fundamental constituents of matter where they can never be "seen."

Recall the conviction with which even the founder of modern science, Galileo, clung to the notion that circular motion is perfect and must therefore apply to the heavens. Without Tycho's observations and Kepler's analysis, would physicists ever have given up this idea? Without observations to agree or disagree with predictions, why should physicists ever adopt a particular vocabulary—or once having adopted it, ever give it up?

On the other hand, proponents of superstring theories may be comforted to know the following words were not directed at them. "The boldest attempt [yet] towards a philosophy of pure idealism [is] . . . merely a logical exercise of the active mind, and ignores the world of brute facts, [it] may be interesting, but it ultimately evaporates into a scholasticism . . . it will cause the decadence of science as surely as the medieval scholasticism preceded the decadence of religion."[75] The provocation for such despair? Einstein's general theory of relativity!

The only way physicists are willing to discriminate between ways of talking about nature is on the basis of the conclusions to which differing vocabularies lead. If two ways of talking make the same predictions, as do Heisenberg's matrix mechanics and Schroedinger's wave mechanics, physicists say that they are dialects of the same language. Furthermore, the appeal of a theory has nothing to do with its effectiveness. No one is particularly fond of the form of the electroweak theory. One of its creators even called the theory repulsive. One of the key steps in developing a theory of the weak interactions was the realization that nature discriminates between right- and left-handedness. The idea that nature should have no such preference is so beguiling that physicists found it extremely difficult to give up the notion until the evidence became overwhelming. It seems safe to say, if physicists could build a world modeled on their heart's desire, it would be a world in which nature was blind to right- or left-handedness. The world does not always fit our hearts' desires. Mathematical elegance, simplicity, and economy are important, but eventually a theory has to pay its way by making accurate predictions. In the words of Steven Weinberg, "We just can't go on doing physics like this without support from experiment."[76]

It is ironic that physicists may have achieved what some of the greatest physicists have pursued—a language unifying all the forces of nature into one model—at the same time that they seem to have exhausted their capabilities for knowing whether this language works or not. But the history of quantum field theories shows us that it takes time to discover the uses of a new vocabulary. It will be some time before physicists will be ready to pass judgment on the latest efforts to create a language in which to talk about the world on what appears, at least for the present, to be the most fundamental level.

The world was simple in the time of the ancient Greeks: then, there were only four elements—earth, air, fire, and water. The world was simple once again in the beginning of this century when there were two kinds of fundamental particles, electrons and protons, and two forces of nature, gravity and electromagnetism. The "simplicity" of the world as physicists see it today involves six kinds of quarks, their associated antiquarks, six leptons and their antiparticles, eight gluons, photons, the W and Z particles, gravitons, and X particles. If supersymmetry is correct, there are twice as many fundamental particles. And in superstring models the world exists in ten dimensions—six of which remain hidden forever in the insides of quarks and leptons—and is characterized by hundreds of fundamental particles. Not yet a very convincing case for the fundamental simplicity of nature.

Many physicists believe the simplicity of nature is revealed by the fact that the world can be described by a family of quantum field theories involving local gauge symmetry and spontaneous symmetry breaking. Listen to Anthony Zee describing how to design a universe:

> Pick your favorite group: Write down the Yang-Mills theory with your group as its local symmetry group; assign quark fields, lepton fields, and Higgs fields to suitable representations; let the symmetry be broken spontaneously. Now watch to see what the symmetry breaks down to. . . . That, essentially, is all there is to it. Anyone can play. To win, one merely has to hit on the choice used by the Greatest Player of all time. The prize? Fame and glory, plus a trip to Stockholm.[77]

Many of the physicists we have met have taken that trip to Stockholm, but is it really because they discerned the choices of the Greatest Player of all time?[78] Might not there be another explanation? Eugene Wigner raises this possibility:

> We cannot know whether a theory formulated in terms of mathematical concepts is uniquely appropriate. We are in a position similar to that of a man who was provided with a bunch of keys and who, having to open several doors in succession, always hit on the right key on the first or second trial. He became skeptical concerning the uniqueness of the coordination of keys and doors.[79]

Feynman suggested the reason for a possible lack of uniqueness:

> The limited imaginations of physicists: when we see a new phenomenon we try to fit it into the framework we already have. . . . It's not because Nature is *really* similar; it's because the physicists have only been able to think of the same damn thing, over and over again.[80]

As the historian of science Gerald Holton has said, human beings sometimes display an incredible viscosity of imagination. When all you have is a hammer, everything sooner or later begins to look invitingly like a nail. Some of the simplicity physicists find in nature may be the result of the procrustean tools (currently the language of quantum field theory) they use to work with nature.

Experiment and Theory

The history of physics is a story of the relationship between experiment and theory. Without the observations of Tycho there could have been no theory of Kepler. Without experiments of Ampere, Oersted, and Faraday there could have been no Maxwellian electromagnetism. The relationship works both ways. Without Einstein's general relativity there would have been no eclipse photographs in 1919 by Eddington. Without Gell-Mann's prediction of the Omega-minus particle, there would have been no search for, and discovery of, the particle by Samios.

Like any close working relationship, the one between theory and experiment is not without tensions. Most experimental physicists would like nothing better than to find something that sends the theoretical physicists back to their offices and computers to develop a new model. Theoretical physicists, on the hand, have learned to view the work of experimental physicists with some perspective, since often a preliminary observation has not stood up. Einstein, for example, ignored the laboratory measurements of Walter Kaufmann, which Kaufmann claimed were incompatible with special relativity. Because of experimental difficulties, it took almost ten years to show Einstein was right and Kaufmann wrong. Likewise, measurements that appeared to conflict with the electroweak theory led many physicists to attempt to construct alternative models, but improved measurements showed that the theory did not need modification after all.

Physics, however, has come to a point where essentially all the experimental results can be explained by existing theory. This does not mean that all questions have been exhausted but rather that physicists are not goaded by unexplained observations or laboratory results to develop new theories—the way Kepler was goaded by Tycho's observations to develop his way of talking about the solar system, or Bohr was goaded by the observations of Fraunhofer and his successors to develop a new way to talk about the atom. The grand unified theories, supersymmetry, and superstring theories in this sense are solutions in search of a problem. With the exception of the as-yet unobserved decay of the proton, these theories have virtually no experimental consequences that physicists can ever hope to see. As a result, they are perilously close to being scholastic exercises rather than science. As grand as they are intellectually, it is difficult to avoid arriving at the conclusion voiced by one of the originators of GUTs, Howard Georgi: "Unification is clearly fundamental, but it may not be physics if you can't see any of the effects."[81]

If the simplest grand unified theories are right, there may be no new physics to be discovered when the next generation of particle accelerator, the proposed Superconducting Super Collider (SSC), is built—other than possibly the Higgs particle.[82] If GUTs are correct, new particles will not be seen until much higher energies than physicists can reasonably ever expect to reach with foreseeable accelerators are reached. (A circular accelerator that reached around the entire earth could produce energetic particles that would fall short of the mark by a factor of 100 million. A linear accelerator capable of reaching unification energies would stretch almost to the nearest star.)

Yet most physicists would be extremely uncomfortable allowing a way of talking about nature to convince anyone that there is no need to make a measurement. Nature has surprised all of us too often. Even the proponents of GUTs support the development of the SSC. Possibly supersymmetric particles or a family of the Higgs particles, will be found. In any case, without new data fundamental physics will have come to a dead end.

Why are physicists so concerned with understanding processes not yet found anywhere in the universe? They believe that the energies of grand unification, and the even higher energies at which general relativity is expected to break down, did occur once—in the earliest fraction of a second of the birth of the universe. The universe may take the form we know as the result of processes that took place during the first fraction of a second of its existence—processes at the unification energy and even higher energies. It may be that physicists will have to turn to the largest possible scale, the universe as a whole, to find the evidence for processes taking place on the smallest imaginable scale.

In the seventeenth century, the German mathematician and philosopher Wilhelm Leibniz, who independently invented the calculus at the same time that Newton did, asked, Why is there something and not nothing at all? Philosophers have puzzled over this question ever since. If one of the grand unified theories proves to fit nature, physicists may be able to answer Leibniz's question—but in a way that would no doubt have surprised him. For the answer would be that "nothing" is unstable.

The energy associated with gravity is opposite in sign to the energy associated with matter. If the universe has just the right amount of matter—and there appears to be close to this amount—the matter will exactly balance the negative gravitational energy, and the net energy in the universe will be zero. But if the net energy of the universe is zero, it could be created out of "nothing" without violating the conservation of energy. The net electric charge in the universe appears to be zero as well.[83] In fact, nothing physicists now know is inconsistent with the net of *all* conserved quantities being zero at a sufficiently high temperature. But if this is true we can talk about the universe being created spontaneously from nothing (physicists call this a *vacuum-state fluctuation*) without violating any of the conservation laws. The universe would then be what the vacuum produces when left to itself! In the words of the American physicist Alan Guth, "The universe could be the ultimate free lunch."[84]

13

The Inscrutable Essence of Mathematics

The miracle of the appropriateness of the language of mathematics for the formulation of the laws of physics is a wonderful gift which we neither understand nor deserve.[85]

EUGENE WIGNER

As far as the propositions of mathematics refer to reality, they are not certain; and as far as they are certain, they do not refer to reality.[86]

ALBERT EINSTEIN

So far we have looked at the way physicists say the world is put together. Bearing in mind that physics is fundamentally mathematical in nature, it might be interesting to see how mathematicians talk about the world. The nineteenth-century mathematician Charles Hermite expressed a belief apparently shared by many mathematicians:

> I believe that the numbers and functions of analysis are not the arbitrary product of our spirits; I believe that they exist outside of us with the same character of necessity as the objects of objective reality; and we find or discover them and study them as do the physicists, chemists and zoologists.[87]

The twentieth-century English mathematician G. H. Hardy wrote:

> I believe that mathematical reality lies outside us, that our function is to discover or *observe* it, and that the theorems which we prove, and which we describe grandiloquently as our "creations" are simply our notes of our observations.[88]

Hardy's view of mathematical reality sounds virtually identical to most physicists' view of physical reality. The twentieth-century French mathematician Jacques Hadamard also put it pointedly: "Although the truth is not known to us, it *pre-exists,* and inescapably imposes on us the path that we must follow."[89] Morris Kline, a present-day mathematician, summarizes this view, apparently widely held by mathematicians: "Thus mathematical truth is discovered and not invented. What evolves is not mathematics but man's knowledge of mathematics."[90]

This point of view raises questions that have been around since the days when Plato first put forward the idea that numbers exist independently of human minds. Mathematical "discovery" seems to be a metaphor. Where is this realm of mathematics? How do mathematicians discover truths about it? How do they know such a realm exists? Or is the truth of mathematics, like so many truths, closely tied to the ways we use language?

Mathematical Conjectures

A prime number is one that can be divided only by one or by itself without leaving a remainder (1, 2, 3, 5, 7, 11, 13, and 17 are examples). In the beginning of the eighteenth century the Russian mathematician, Christian Goldbach conjectured that every even number larger than two can be written as the sum of two primes. This conjecture is obviously true for the smallest even numbers: $4 = 3 + 1$, $6 = 5 + 1$, $8 = 7 + 1$, $10 = 7 + 3$, $12 = 11 + 1$, $14 = 13 + 1$, $16 = 13 + 3$. . . . In fact, Goldbach's conjecture has been shown to be true for a very large number of even numbers. There is, however, no proof that Goldbach's conjecture is true for *every* even number, which is, of course, why it is a conjecture. Yet many mathematicians would say that even though the statement "Goldbach's conjecture is true" may never be proved, it *must* be either true or false. Either there is an even number in the realm of pure numbers that is not the sum of two primes, or there is no such number. Does this argument have a familiar ring? It sounds somewhat like the argument that the electron *must* pass through one slit or the other. We know this argument is wrong for electrons. Could the argument be wrong for Goldbach's conjecture as well?

We have a strong penchant for maintaining that a statement must be either true or false. Indeed, there is a principle in logic, called the *law of the excluded middle*, or *bivalence*, which says that every statement must be either true or false. But how do we know the law of the excluded middle is true? In fact, of course, we don't. The law is a convention—a way we use language.

172

Compare Goldbach's conjecture with another conjecture—the conjecture that a woman never will be elected president of the United States. Is this conjecture true or false? If we insist it must be one or the other, we seem to be committing ourselves to a future that somehow already exists, for the truth or falsity of the statement depends on events that have not yet occurred. Why not say this new conjecture is neither true nor false? And if the new conjecture is neither true nor false, why not Goldbach's conjecture? In other words, what do we really gain by saying that every statement must be either true or false? Maybe mathematicians will never find a counter example to Goldbach's conjecture; maybe it will always remain a conjecture—neither true nor false but something else. Goldbach's conjecture, after all, is something mathematicians *say,* just as the law of the excluded middle is something logicians *say.* Maybe the realm of pure numbers is language. Maybe mathematicians explore what they can consistently say.

In view of these possibilities, it should come as no surprise that there are mathematicians who dissent from the majority view. For example, the nineteenth century German mathematician Richard Dedekind wrote, "Number [is] not the class itself, but something new . . . which the mind creates. We are of a divine race and we possess . . . the power to create." His fellow German Karl Weierstrass said, "The true mathematician is a poet." Percy Bridgman, a twentieth-century American Nobel laureate in physics, said in characteristically no-nonsense terms, "It is the merest truism, evident at once to unsophisticated observation, that mathematics is a human invention."[91]

The great twentieth-century mathematician Kurt Gödel is often thought of as belonging to the group that maintains that mathematics describes a realm independent of the human. However, as we have seen physics develop, it is possible to give a somewhat different reading to his words about the nature of mathematical sets:

> It seems to me that the assumptions of such objects is quite as legitimate as the assumption of physical objects and there is quite as much reason to believe in their existence. They are in the same sense necessary to obtain a satisfactory theory of mathematics as physical bodies are necessary for a satisfactory theory of our sense perceptions.[92]

After tracing the development of physics, we might well agree that mathematical objects are as real as physical objects, without being quite certain about exactly what that tells us. Mathematics is clearly a language, but does a language have to describe something outside of language? Despite the abstract nature of mathematics, most mathematicians seem convinced of the objective reality of the world of which they write. But does this tell us something about the world or something about the way we speak? Perhaps the distinction between creating and discovering is not as sharp as it seems. Did Einstein discover relativity or did he invent it? Did Gell-Mann discover quarks or did he invent them?[93] Did the sixteenth-century Italian mathematician Raphael Bombelli discover imaginary numbers or did he invent them?

It seems difficult to deal day in and day out with something, no matter how abstract, without developing the conviction that what we are working with is real—is independent of what we do and say about it. This conviction provides us with a clue to the nature of the "furniture of the universe," the stuff we are convinced the world is made of. As far as we are concerned, ontology recapitulates taxonomy—the way we divide the world in language tells us how we think the world is "really" put together.[94]

We have been talking about models of nature. Mathematicians also talk about models, but they often use the word differently from physicists. A mathematical model is an interpretation of a mathematical system. What exactly does this mean? One way to answer this question is to take a brief look at geometry.

An Unparalleled Effort

Since the time of Euclid, geometry has represented a model of rational thought. Beginning with a series of clear definitions and self-evident axioms, Euclid's *Elements* developed a series of propositions whose certainty was indisputable. Many of the greatest philosophers, scientists, and mathematicians turned to Euclid's example in the effort to develop an equally fundamental level of certainty.

Since the beginning of geometry, however, one of Euclid's axioms seemed less compelling than the others—the fifth axiom, or as it is often called, the parallel postulate. In its modern form, the fifth axiom states: Through any point not on a given line, one and only one line can be drawn parallel to the first line. Although no one doubted the truth of this axiom, it seemed somehow less immediately obvious than the other axioms. Some of the greatest mathematicians, starting with Euclid and including the great German mathematician Karl Friedrich Gauss at the end of the eighteenth century, tried to demonstrate that the parallel axiom could be derived from the other axioms. All these attempts failed.

Gauss, in fact, became convinced that it was possible to develop a logical geometry in which the parallel axiom did *not* hold. He reasoned that if the parallel axiom cannot be derived from the other Euclidean axioms, then changing the parallel axiom will not conflict with the other axioms. Gauss's friend Johann Bolyai, and, independently, the Russian mathematician Nickolai Lobatchevsky, developed alternative versions of geometry in which the parallel postulate is replaced by the assumption that *more* than one parallel can be drawn through a point not on a given line.

Life became even more complex with the development of yet another non-Euclidean geometry. Gauss's student, Georg Riemann, altered the parallel postulate in a way different from the way Gauss, Bolyai, and Lobatchevsky did. Riemann postulated that *no* parallel lines can be drawn through a point not on a given line. In Riemann's geometry all straight lines have the same length, and each pair of lines meet at two points. How could this bizarre system be called geometry?

The Italian mathematician Eugenio Beltrami showed one way to understand Riemannian geometry. But to appreciate Beltrami's contribution, we must look at how we determine whether a line is straight. You may recall this point was first raised in the discussion of Einstein's general theory of relativity.

Over short distances drawing a straight line does not seem to pose any problem. Surveyors seem to have little difficulty specifying what they mean by a straight line. But what happens when we deal with distances on a global scale? What does a straight line drawn between New York and Tokyo look like? Here the answer is not so obvious.

The earth is spherical, and we can draw many lines between Tokyo and New York. As long as we keep on the surface of the earth, all of them look curved. How can we call any of them straight?

Perhaps we should pose another question. What route is the shortest distance between New York and Tokyo? Here there *is* a unique answer. The shortest air distance between New York and Tokyo lies on a circle passing through both cities and near Alaska. The center of this circle lies at the center of the earth. The plane defined by such a circle divides the earth in half and for this reason is called a *great circle*. On most flat maps of the earth a great-circle route that is not along the equator is usually far from the shortest distance between two points. On a flat Mercator map, the kind we normally see, the great circle route from New York to Tokyo seems to stray far from the "direct" path. But flying by way of Alaska minimizes the time it takes to travel between the two cities—the Alaskan route is the shortest route.

We have no problem saying that on a piece of paper a straight line is the shortest distance between two points. If we allow the same definition of a straight line to apply on the surface of the earth, we find that a great circle is a "straight" line. Beltrami pointed out that if we *call* a great circle a straight line, then Riemannian geometry is the geometry of the surface of a sphere. Since any two great circles are the same length and intersect twice, they fulfill Riemann's version of the parallel postulate which says any two straight lines always meet.

Beltrami provided what mathematicians call a model of Riemannian geometry. Recall that for a mathematician, a model is a way of interpreting a mathematical system. For example, Euclid's axioms seem clear enough. We think we know what a point and a line are. But as Beltrami showed us, by slightly altering what we call straight, we can create a different model of the axioms. Furthermore, we can go even further than Beltrami did.

Mathematical Models

Imagine we want to construct a set of axioms to allow us to develop theorems that are true with respect to bees. The first axiom might be, "Each bee belongs to one and only one hive." A second axiom might be, "There is one and only one queen bee in a hive," and a third axiom, "The total number of bees in a hive is equal to the number you start out with plus the number that are born minus the number that die."

From these axioms we could prove a number of things about beehives, such as "If x is a queen bee, and y is a queen bee, and if x and y are in the same hive, then x and y are the same bee." But suppose instead of interpreting the axioms this way, we were to interpret *bee* to mean "book," *hive* to mean "library," and *queen bee* to mean "librarian." Now our axioms, instead of being about bees, are about libraries. For example, the first axiom now says, "Each book belongs in one and only one library," and the third axiom, "The total number of books in a library is the number you start out with plus the number you buy minus the number that are out on loan." Furthermore, any theorem we proved about beehives now proves to be just as true of libraries. A mathematician would say that beehives and libraries are two models of the set of axioms we developed. Both models are equally right as far as a mathematician is concerned, and anything we prove from the axioms will hold equally well for either model.

By showing the geometry of a sphere to be a model of Riemannian geometry, Beltrami showed that if Euclidean geometry is a consistent system, so is Riemannian geometry. Thus Beltrami showed that Euclidean geometry is not a unique set of truths. Rather, it is important because the world we encounter every day, the local scale on which the earth seems quite flat, is a good model of the Euclidean axioms.

Think back to our example again. Suppose instead of the axioms applying to a universe of bees, we ask if they apply to a universe consisting of bees *and* the Empire State Building. Sure enough, we find they do. In other words, there seems to be no limit to the number of models that can be developed for a set of axioms.

Someone may argue that a universe of bees alone is simpler than a universe of bees and the Empire State Building and therefore the example is misleading. But is this as obvious as it seems? Is a universe with imaginary numbers simpler than a universe without them? Is a universe with Higgs particles simpler than a universe without them?

Mathematicians think of a mathematical system as a formal language that can be interpreted by providing a model of the system. For a physicist a mathematical system is a model of the physical world. For each of them an abstract language can be mapped onto features of the world in a variety of ways. But there seems to be no unique way of combining a mathematical language and a part of the world. Our words apparently do not hook onto the world in only one way. If there is no absolute relationship between physical theories and the world, how can we know how the world is "really" put together?

14

The Unspeakable Power of Language

The sense experiences are the given subject matter. But the theory that shall interpret them is man-made. It is the result of an extremely laborious process of adaptation: hypothetical, never completely final, always subject to question and doubt.[95]

ALBERT EINSTEIN

Theories and Reality

Classical physics demonstrated the power of a language that separates the observer and the observed, the subject and the object. Surely there are facts about the way the world is independent of what we say, and surely we can talk about the world as it is independent of any observations. Quantum mechanics, however, did not fit this framework. The success of quantum mechanics showed physicists that when they talk about the atomic realm, they can no longer talk of a world whose behavior can be described in the absence of a well-defined scheme of measurement. In talking about the atomic world, the observed and the observer cannot be separated the way they can be when we talk about the world of everyday experience. To go beyond the realm of classical physics, physicists had to give up the paradigm of a detached observer and an independent reality.

In the final analysis, physics is only indirectly about the world of nature. Directly, it is talk about experimental arrangements and observations. Given a particular experimental arrangement, physicists can predict the outcome of certain measurements. There is nothing arbitrary about these outcomes. Anyone with the requisite ability can replicate them—they are perfectly objective in this sense. Nor is there anything arbitrary about the predictions. What is not given to physicists by nature, but rather is invented by them, is what they *say* about these outcomes, the language they use to talk about nature. If physicists try to step outside the scheme of experimental arrangements and observations to envision what sort of independent mechanism in the world "really" produces those observations, in Feynman's words, they "get 'down the drain', into a blind alley from which nobody has yet escaped."

How do we know what the world is like? We think our eyes give us a representation of the way the external world really is, but as the prologue points out in the example of the photograph of a building, this too is a way of speaking—a way of speaking based on the notion of "realistic" representation. In fact, theories have been developed that do not endow the nervous system with a representational function. In a recent book describing vision and illusions of color, the biologists Humberto Maturana and Francisco Varela say:

> Because these states of neuronal activity (as when we see green) can be triggered by a number of different light perturbations (like those which make it possible to see color shadows), we can correlate our naming of colors with states of neuronal activity but not with wavelengths. What states of neuronal activity are triggered by the different perturbations is determined in each person by his or her individual structure and not by the features of the perturbing agent. . . . Doubtless . . . we are experiencing a world. But when we examine more closely how we get to know this world, we invariably find that we cannot separate our history of actions—biological and social—from how this world appears to us."[96]

Perhaps we should not be too surprised if we cannot grasp an absolutely independent world. To paraphrase Charles Darwin, we are organisms shaped, not by getting the world right, but by surviving to leave offspring. Before we embrace the idea that survival is invariably aided by getting the world right, that is by representing the world "correctly" in language, perhaps we should look at the living things that have survived and left offspring for hundreds and even thousands of times longer than *Homo sapiens*. The frog, for example, sees a very different world from the one we see. If frogs had language, it would be illuminating to learn what they would say about how the world is "really" put together. The perception of a causal world unfolding in space and time, which serves our survival so well, may be only a tool that works on the scale on which it evolved.

Some say quantum mechanics shows that experimental arrangements compel electrons to take on certain values such as position and momentum. But we can equally well say that there are no facts about the paths of subatomic particles; instead, there are our interpretations of the measurements we make—interpretations in terms of position and motion. Physicists discovered that they cannot interpret their measurements in a language where position and momentum are simultaneously precise.

The existence of a world we cannot see makes sense from a physicist's point of view only if this world has observable consequences. Physicists cannot "see" quarks or gluons, but quarks and gluons are elements of physical theory because they lead to predictions that physicists *can* see. Talking as though there are quarks and gluons helps physicists to make sense of the world.

Knowledge and Reality

There is a sense in which no one, including philosophers, doubts the existence of a real objective world. The stubbornly physical nature of the world we encounter every day is obvious. *The minute we begin to talk about this world, however, it somehow becomes transformed into another world, an interpreted world, a world delimited by language*—a world of trees, houses, cars, quarks, and leptons. In order to deal with the world we have to talk about it (or measure it, or shape it—in any case we engage the world in terms of our symbols, whether we are building a pyramid or a Superconducting Super Collider).

When people first talked about electrons, they thought of them as perfectly respectable bits of matter—too small to be seen directly but otherwise no different from objects on the scale we are familiar with. Einstein's argument that matter and energy somehow must be equivalent suggests that all is not so simple as it might at first seem; Dirac found we can talk of electrons and positrons as being created out of nothing but energy and being reduced to nothing but energy. Quantum mechanics shows it is impossible even to picture these elementary constituents of matter: They are required to be well-behaved localized entities whenever we detect them, but otherwise diffuse "possibilities of detection" spread widely over space and time.

Quantum electrodynamics replaces discrete electrons with talk about excitations of an electron field: a field that, in addition to harboring the probability of observing "normal" electrons, teems with virtual electrons and positrons winking into and out of existence. In superstring theory, electrons are the energy states of incredibly tiny quantum strings vibrating in ten-dimensional spacetime. These descriptions are different ways of talking about the same world. The best way to understand the role of a theory seems to be as a tool, a way of speaking, appropriate or inappropriate to the task at hand.

It seems perfectly reasonable to ask whether leptons and quarks are two kinds of "stuff" out of which the world is made. Mathematicians, however, talk about numbers in much the same way that physicists talk about leptons and quarks. A physicist's world is made up of leptons and quarks because physicists talk about their experiments in terms of leptons and quarks. In the same way, the mathematician's world is made up of imaginary numbers and infinite sets because imaginary numbers and infinite sets are an essential feature of the discussions of contemporary mathematics; an economist's world includes markets, supply, and demand for a similar reason. The word *real* does not seem to be a descriptive term. It seems to be an honorific term that we bestow on our most cherished beliefs—our most treasured ways of speaking.

The lesson we can draw from the history of physics is that as far as we are concerned, *what is real is what we regularly talk about*. For better or for worse, there is little evidence that we have any idea of what reality looks like from some absolute point of view. We only know what the world looks like from *our* point of view. From a physicist's perspective, the behavior of the physical world is most effectively talked about in the language of quantum field theories.

The Conversation of Physics

If the current way of talking about quantum chromodynamics continues, physicists will never see an individual quark or measure its fractional electric charge. The effects of the color force will forever remain invisible. Nature seems to conspire to keep physicists from ever seeing her ultimate constituents—or, more accurately, the most successful ways of talking about nature that physicists have found turn out to require that they speak in terms of fundamentally unobservable elements. Yet most physicists are committed to the reality of quarks. It is hard to imagine working every day with an idea without being committed to its reality. As Einstein said, "Without the belief that it is possible to grasp reality with our theoretical constructions, without the belief in the inner harmony of the world, there would be no science."[97]

The role of language is always easier to see when someone else's language is involved. While physicists talk about quarks the same way they talk about anything else, those who are not working physicists have the luxury of stepping back and seeing that quarks are a way of talking about the world—a way of talking that gives physicists power in describing and predicting nature's behavior.

In Heisenberg's words, "What we observe is not nature in itself, but nature exposed to our method of questioning. Our scientific work in physics consists in asking questions about nature in the language we possess and trying to get an answer from experiment by the means that are at our disposal."[98] And Bohr's, "It is wrong to think that the task of physics is to find out how nature is. Physics concerns only what we can *say* about nature."

We have always dreamed of being able to talk about the world in the *right* way—to talk about the world in what the American philosopher Richard Rorty ironically calls "nature's own language."[99] The history of physics makes it hard to sustain the idea that we are getting closer to speaking "nature's own language." Talk about quarks arises in the interaction between physicists and the world; we interact with the world and create interpretations of what this interaction means.

Truth as Procedure

Just as many mathematicians talk as though every statement were either true or false, we human beings want to talk about the physical world in the same way. Either matter is made up of quarks or it is not. Either there is intelligent life outside the solar system or there is not. Either 1978 was the snowiest winter in Boston in this century or it was not. We want to believe there is some unique way our words hook onto the world and that this hooking, or accurate representation, makes our statements true or false. Relativity and quantum mechanics make it hard to maintain the convention of an absolute word-to-world fit. For example, relativity, much to the discomfort of some people, shows that the truth of whose clock is running slower, mine or yours, depends on the frame of reference from which the statement is made.

Quantum mechanics takes us further from the classical worldview. We simply cannot say, "Either the electron goes through the first slit or it doesn't" if we do not arrange an experimental apparatus so that we can test this statement. Furthermore, arranging apparatus in this way precludes displaying other phenomena such as interference.

In Einstein's words, "This universe of ideas is just as little independent of the nature of our experiences as clothes are of the form of the human body."[100] What we say about the world, our theories, are like garments—they fit the world to a greater or lesser degree, but none fit perfectly, and none are right for every occasion. There seems to be no already-made world, waiting to be discovered. The fabric of nature, like all fabrics, is woven by human beings for human purposes.

What does this say about truth? If there can be innumerable theories of the world, how can there be a unique relationship between the world and our theories, between the world and what we say about it? In the conversation of physics, truth is largely procedural. Physicists are united by a procedure that allows them to determine the value of a theory. At some point this procedure involves an appeal to observations that others are free to make as well. The struggle between Galileo and the Church was a struggle over the procedure to be used in determining the truth of certain statements. The resolution of that conflict consisted in distinguishing two domains of inquiry, religion and science, each with its own procedure for determining truth. In the realm of most religions the appeal is to authority; in the realm of science the appeal is to observations and experiments.

Knowledge of the natural world has been advanced by an international community of scientists engaged in a common conversation but one carried out under different rules than most conversations. As Samuel Ting said, "Science is one of the few areas of human life where the majority does not rule."[101] What scientists have in common is not that they agree on the same theories, or even that they always agree on the same facts, but that they agree on the procedures to be followed in testing theories and establishing facts. Physics is primarily procedural. Its procedure is to uncover the value of a theory by determining its consequences and then seeing if these predictions are confirmed by measurements. A physical theory must make predictions that can either agree with or conflict with observations. Kepler, for example, abandoned the idea of circular orbits because predictions made on the basis of circular orbits conflicted with Tycho's observations. If a theory has no consequences that might possibly clash with observation, the theory is not a *physical* theory but some other sort of theory—aesthetic, religious, or philosophical.[102]

The language of general relativity replaced the language of Newtonian gravity in some conversations, because someone developed a procedure, the eclipse photographs, that allowed the value of the two ways of speaking to be assessed by bringing their differing predictions into confrontation with measurement. Once this confrontation occurred the triumph of relativity seemed assured. No matter that the overwhelming majority of physicists were quite convinced that Newton was "right"; Eddington's observations showed them that Einstein's way of talking about gravity was a better way of predicting the outcome of certain experiments than Newton's. Once Eddington had published the results of his measurements, every other physicist could know the power of Einstein's language as well.

The value of a theory is *not* that it fits what physicists already know but that it points to what they do not know. General relativity was embraced because it predicted an as-yet unobserved deflection of starlight. The eightfold way was accepted, not because it provided a rationale for what was already known, but because it predicted something *not* already known, the Omega-minus particle—and this prediction was supported by experiment.

To say scientific truth is procedural is not to say that there is a unique procedure. In fact, there are many procedures. The point is that the same procedure can be used by anyone else who has the same questions. If we want to know whether a particular way of talking about the physical world is valuable, we look for the predictions the theory makes and compare these with observations.

The measurements of physics must be repeatable by other physicists in other laboratories.[103] This repeatability, not the agreement of physicists or the accurate picturing of nature, makes physics both public and "objective."

The Unreasonable Success of Physics

If physics is a conversation, and the search for truth procedural, how can we account for its amazing success? Why can't we simply say a theory works because it accurately corresponds to nature in some way—because the language of the theory represents the way the world "really" is? Unfortunately, to say that nature corresponds to our theories is no more informative than the explanation provided by Molière's fictional physician who reassured his patients that morphine works because of its "dormitive powers." The problem with a dormitive powers theory is that it does not tell us any more than we already knew—morphine induces sleep.

Mathematics, as Wigner says, is unreasonably effective in describing the physical world; it is unreasonable precisely because we can give nothing that would count as a reason. When Newton's approach failed to lead to an accurate description of the detailed behavior of Mercury's path around the sun, there was no way to explain its failure in Newtonian language. Such an explanation had to await the development of a new vocabulary, Einstein's general relativity, that could accurately describe Mercury's behavior. As far as physics is concerned, reasons exist *within* the framework of a theory, not outside it. For example, it makes perfect sense to say that a particular model, such as the description of the roller coaster, works because friction can be ignored or because gravity is a conservative field. Newtonian mechanics provides the framework within which these words function as explanations.

Explanations that appeal to principles outside a theory tend to be uninformative. It does not help much for me to say that the roller-coaster model works because of the additive properties of the real number system, and it helps not at all to say the model works because it corresponds to nature. A way of talking about the world either works in some particular situation or it does not, but we add nothing to our stock of knowledge by saying its success or failure is because of correspondence or lack of correspondence to the world.

How then can we account for the power of Newton's laws, Maxwell's equations, relativity and quantum mechanics? How can it be that a few equations are capable of describing the behavior of systems ranging in size from the atom to the observable universe? Perhaps, as Weinberg tells us, "The standard model works so well simply because all the terms which could make it look different are . . . extremely small."[104]

Einstein addressed the question more broadly. He said, "The eternal mystery of the world is its comprehensibility."[105] This comprehensibility, our ability to talk about the world in the language of mathematics, is the blessing that Wigner points out we neither understand nor deserve. To try to understand, to try to solve Einstein's mystery and uncover the source of Wigner's blessing, is to leave the realm of physics and to enter the realm of metaphysics. A fascinating trip, no doubt, but a journey on which physics has little to offer by way of illumination. As the philosopher Ludwig Wittgenstein said, "What we cannot speak about, we must pass over in silence."[106]

Einstein described the nature of physics in the following way:

Physical concepts are free creations of the human mind, and are not, however it may seem, uniquely determined by the external world. In our endeavor to understand reality we are somewhat like a man trying to understand the mechanism of a closed watch. He sees the face and the moving hands, even hears it ticking, but he has no way of opening the case. If he is ingenious he may form some picture of the mechanism which could be responsible for all the things he observes, but he may never be quite sure his picture is the only one which could explain his observations. He will never be able to compare his picture with the real mechanism and *he cannot even imagine the possibility of the meaning of such a comparison* [my italics].[107]

Einstein said we cannot compare our theories with the real world. We can compare *predictions* from our theory with *observations* of the world, but we "cannot even imagine . . . the meaning of" comparing our theories with reality.

The Archimedean Perspective

Archimedes was perhaps the greatest of ancient Greek scientists. Flushed with having developed the mathematical principle of the lever, Archimedes is said to have exclaimed, "Give me a place to stand, and I will move the earth." The idea of a place outside the world on which to stand has become a fundamental myth of our culture. This myth has most often taken the form of a spectator view of knowledge—the notion we can stand aside from the action and comment on it from a detached viewpoint. The epitome of the spectator view of knowledge was expressed by the eighteenth-century French astronomer Pierre-Simon Laplace:

> We may regard the present state of the universe as the effect of its past and the cause of its future. An intellect which at any given moment knew all of the forces that animate nature and the mutual positions of the beings that compose it, if this intellect were vast enough to submit the data to analysis, could condense into a single formula the movement of the greatest bodies of the universe and that of the lightest atom: for such an intellect nothing could be uncertain; and the future just like the past would be present before its eyes.[108]

What could be more detached than the spectator who views the entire universe, past, present, and future, as an object of contemplation? (Needless to say, for Laplace the scientist has the perspective of the vast intellect he describes, if only to a limited extent.) The spectator perspective is an integral part of the deterministic world portrayed by classical physics.

The idea that we can step outside our systems of interpretation, our language, and somehow talk sensibly about the world as it really is seems to be one of the most deep-seated beliefs we have. The concept reached its scientific zenith in the triumphs of classical physics. The English physicist and cleric J. C. Polkinghorne said, "Classical physics is played out before an all-seeing eye."[109] The "all-seeing eye" is a metaphor, so much so that it never appears to us to be an assumption, but seems simply to be "the way things are." It is this metaphor, acting as an undisclosed assumption that allows us to talk about how our theories reflect the way the world really is. That is, it allows us to talk about comparing theories with the world. The metaphor of an all-seeing eye entrances us. As Wittgenstein said, "A *picture* held us captive. And we could not get outside of it, for it lay in our language, and language seemed to repeat it to us inexorably."[110]

Languages and Metalanguages

How could we compare a theory with the world? A physicist's theory is a collection of mathematical formulas, and the world . . . well, the world is something completely different. Then how can a theory refer to the world? If I say, "The dog is named Willard," I am surely using the word *Willard* to refer to a real animal. There is obviously a difference between the word *Willard* and the dog. When I say, "Electricity is related to electrons in the wire," am I not doing exactly the same thing? Am I not using the word *electrons* to refer to something out there in the world?

In order to talk about the relationship between a theory and the world, we can make use of what a logician calls a *metalanguage*. A metalanguage is a way of talking about a language while using language and, we hope, not becoming too confused in the process. For example, if I say, "The French word for dog is *chien*," I am using English as a metalanguage to talk about words in the French language.

Let us look at the statement, "The word *electron* refers to a subatomic particle." In this case, the language and the metalanguage are both English. We can discriminate between the words we are talking about and the words we are using to do the talking by putting the former in italics. The sentence in the metalanguage tells us the word *electron* refers to something in the world called a subatomic particle. But notice that reference is a relationship between the word *electron* and the words *subatomic particle,* not between words and something that is not words. The observations with which physicists compare their predictions are not some mute expression of the world. They are symbolic and gain their meaning and value in a system of interpretation. No experiment, in Born's words, has any meaning at all until it is interpreted by a theory. Then how do we ever manage to talk about the world?

If you ask me to bring you a book, you judge the success of your request by the arrival of the book. If I bring you the wrong book, you are not likely to feel that there was some failure of your words to correspond to the world but rather that you were not specific enough or that I did not know which book you meant. No discussion of the nature of reference would improve the situation, and a discussion of the reality of books would be even less likely to be helpful. Books are real, not because of some mystical connection between language and the world, but because you can ask me to bring you a book and my action can fulfill your expectation. That expectation and its fulfillment are made possible by our community of shared assumptions, conventions, and understandings—our shared language. Unicorns are not "real" because our community has no expectations about living or dead unicorns that can be fulfilled, only mythological ones. The reality of unicorns is related to our use of the word *unicorn*. Using the word every day would commit us to the existence of "real" unicorns, just as using the word *quark* every day commits physicists to the existence of quarks.

If we question the existence of certain things, the luminiferous ether and unicorns, for example, we question whether these words continue to pay their way in our speech. If *luminiferous ether* and *unicorn* fall into disuse, the ether and unicorns are no longer real. But notice that *we* are sure that the ether and unicorns were *never* real. When we stopped using these words, our new vocabulary seemingly reached back into the past and eliminated any reality the luminiferous ether and unicorns *ever* had. History is not as immutable as we might think; language can apparently transform the past as readily as it shapes the present and the future.

15

The Last Word

It is the theory that decides what we can observe. [111]
ALBERT EINSTEIN

An electron is no more (and no less) hypothetical than a star. [112]
SIR ARTHUR STANLEY EDDINGTON

Three umpires were discussing their roles in the game of baseball. The first umpire asserted, "I calls 'em the way I sees 'em." The next umpire, with even more confidence, and a more metaphysical turn of mind said, "I calls 'em the way they *are*!" But the third umpire, displaying a familiarity with twentieth-century physics, concluded the discussion with, "They ain't *nothin'* until I calls 'em!" The ball, the bat, and the plate do not create the game; the rules create the game, and the umpire interprets the rules, and in the process, creates the score. The players and the fans have no doubt that the ball was either over the plate or not over the plate, but the umpire's call, and not any fact of the matter, creates a ball or a strike.

The Role of Language

Implicit in the way we use language is the notion that language points to a world beyond itself. In the everyday world, language points to trees, buildings, automobiles, cabbages, and kings. But what kind of stuff is this everyday world made of?

Surely the world is not *really* made up of supply and demand, profit and loss, or democracy and communism. These are human ways of characterizing experience; we want to know what the world is like independent of human concepts. We want to know whether the runner was really safe or out. If we want to know what the real substance of the world is, we have to turn to the hard sciences, and physics is the hardest science of all. If any science can claim to know something about the way the world is put together, physics can. So we have turned to physicists to find what the world is really made of, and they have told us. But their answers have hardly reassured those of us looking for certainty. In our attempt to get to a world outside of language, we have apparently wound up squarely in the net of language. When someone argued with Niels Bohr that reality is more fundamental than language, he responded, "We are suspended in language in such a way that we cannot say what is up and what is down. The word 'reality' is also a word, a word we must learn to use correctly."[113]

At the heart of physics is the process of building models of the world. Often these models have required the invention of new mathematical languages. The models physicists build can never be compared with the world, because as Einstein tells us, we do not even know what that would mean. Rather, the predictions derived from the models can be compared with observations of the world. However, this comparison is by no means simple and straightforward. For example, only by a chain of complex reasoning can physicists link the tracks in a bubble chamber with the existence of an invisible neutrino or equally invisible quarks.

The hallmark of the conversation of science is the willingness to allow the agreement or disagreement between predictions and observations to determine whether physicists are satisfied with the theory they have or whether they try to build a new theory—that is, whether they continue to talk about the world in one way or change to another. In physics, this fidelity to observation started with Kepler, who threw away years of work and centuries of conviction because his theory's predictions differed from Tycho's observations by a fraction of a degree. The importance of observation continues to the present day, when a minute discrepancy between the observed and predicted energies associated with the configuration of the electrons in hydrogen atoms spurred the development of quantum field theory.

To a physicist the world is comprehensible to the extent that the world's behavior is predictable. In this sense, the ancient astrologer's drive to predict the future survives in the modern physicist. The difference between physics and most human enterprises is the physicist's ability to use physical theories to make predictions successfully. Despite the uncanny success of mathematical models in making such predictions, the richness of nature far exceeds physicists' ability to capture nature in a mathematical simplification. Once physicists pass beyond the simplest of systems, they find it incredibly difficult to make predictions by solving the mathematical equations describing the detailed behavior of the world. Instead, they are forced to work with approximations and averages that sometimes allow them to predict quite well how things will turn out and sometimes lead to abysmal failures. Physics is the drive to predict the behavior of a world stripped of most of its complications. Chemistry, with its profusion of atoms and molecules interacting in intricate ways, is more complex than physics, and biology is immensely complex.

The language of physics is well suited to talking about the world at the most elementary level we can imagine, but it would be silly to try to use the vocabulary of quarks and leptons to talk about living things or social systems. Physics can perfectly well explain that sunlight scattered by the molecules of gas in the atmosphere produces the blue color of the sky, but attempting to talk about the beauty of the sky in the language of quantum electrodynamics seems hopeless: "Physicists may one day have found the answers to all physical questions, but not all questions are physical questions."[114] Even in describing complex physical situations, however, physicists are like novice chess players who find themselves playing an international grand master. Knowing the rules of the game hardly guarantees being able to play a credible game—much less emerging victorious.

My dog can look in the direction of the sunset, but I somehow doubt the sunset holds any beauty for her or that she even sees a sunset. There is more to beauty than meets the eye. Beauty seems to lie, less in the eye, and more in the mouth, in the language, of the beholder. It is hard to argue that there is a fact about beauty aside from our conventions and our language. Despite our saying that the reaction we feel to beauty is a matter of feeling or emotion, the beauty of the sunset seems to grow out of the distinctions language makes possible. In the same way, the beauty the physicist sees in the heart of nature is a result of the distinctions made possible by the language of physics.

For thousands of years questions have been raised about the nature of reality. Is reality material or is it spiritual? Is reality one or is it many? Is everything determined or is there room for freedom? The answers have not proved persuasive, since the questions are still being debated. Physics shows us a different way to look at these questions—a way that asks what vocabulary, what theory, we should use to talk about the world. The word *should* makes sense in terms of the ends we hope to achieve. The question need not be whether reality is material or spiritual; it can be, what follows from talking about reality in one way or the other? What do we gain, and what price do we pay, for adopting one vocabulary and giving up another?

Like the Greek gods on Mount Olympus, the quarks and leptons of the present day are a story. How long physicists will continue to tell the story outlined in this book depends on how long it serves a useful purpose. Quarks and leptons are a *façon de parler*. They are no different in this respect from houses, trees, or stars. Language tells us what the world is made of, not because language somehow accurately captures a world independent of language, but because it is the heart of our way of dealing with the world. When we create a new way of talking about the world, we virtually create a new world. This observation is no more profound, nor any less profound, than saying that the questions we ask determine, not the content of the answers we will get, but what will *count* as an answer. When a pickpocket encounters a saint, all the pickpocket sees are the saint's pockets.

Just because all we have are stories does not mean that all the stories we have are equal. A story does not have intrinsic value; it has value only to the extent that it serves a purpose, and the stories of physics have served their purposes very well indeed. Physicists have gained enormous predictive power by letting the agreement between predictions and observations tell them which ways of talking to keep, and which to discard.

Explanations, no matter how wonderful, are stories about how we got from where we were to where we are. Clearly some explanations are more compelling than others. But the history of physics shows us that we lack the ability to judge whether an explanation is powerful unless we can tell what its consequences are. Explanations that cannot predict—and predict accurately—are like Kipling's just-so stories, in Heisenberg's words, "matters of personal belief." Viewed from this perspective, *most* of our explanations are just-so stories. They function in various ways, including amusing or annoying us, alienating us or building solidarity. It is hard, however, to maintain that these stories succeed in doing what physics has not been able to do—to picture the world as it somehow really is rather than as it seems to us to be. We might be better off regarding our most cherished beliefs as ways of talking about the world, rather than revealed truths.

Quantum mechanics shows us that there are no facts with regard to the precise paths of electrons. While it is true that what we say determines what kinds of things will turn out to be facts in the matter, we are not free to say anything we like—at least not if we want to talk physics. For, as we have seen, the world sharply constrains the kind of things physicists can say and still make accurate predictions.

Outside of the structure provided by a language, it does not make much sense to say there are "really" facts. Quarks and leptons are interpretations of mathematical expressions. A similar statement applies to forces and quantum fields. What can count as a fact is determined by our language, not by the world. In Einstein's words, "It is the theory that decides what we can observe." The German philosopher Martin Heidegger put the matter more poetically, "Language is the house of being."[115] The language we use tells us the kind of a world we can expect to find.

If, as Einstein told us, there is no ultimate theory, no ultimate language, then there is no ultimate fact about the stuff the world is made of. But if this is so, how much sense does it make to insist that there is a fact, apart from the language we use, with regard to anything else? Physics has a clear criterion for the choice of a language—physicists choose the language that allows their predictions most closely to fit their observations. But outside of physics, outside of science, what criteria do we employ to determine the vocabulary we will use and the "facts" we will therefore find?

Watching the changing way physicists talk about electrons, for example, suggests that the notion we seem to have that names are like labels on museum displays does not hold up very well. In the world of physics, names seem more like the descriptions of animals that are free to roam in a modern zoo. Furthermore, the descriptions of these occupants were not written by an omniscient curator, but rather jotted down by earlier visitors, perhaps describing something only glimpsed in the distance. It is sometimes hard to tell exactly what these phrases describe or if we are looking at exactly the same animals the earlier visitors saw.

The world is not infinitely fluid. Physicists today can perform the same experiments with cathode-ray tubes that Thomson first performed almost a hundred years ago. They will see similar effects and be able to describe them almost exactly as he did. Thomson would also have relatively little difficulty understanding the principles behind a modern television set. What would be very different, however, is what Thomson and a modern physicist would *say* about the observations they make—the language they would use to interpret their findings. It is in this sense that modern physicists live in a world different from Thomson's.

Reference is a word, and like any word it has to be used carefully. Our language commits us to the existence of house, dogs, fire engines, and quarks. When we finally settle down into the language that we *use* to get our day-to-day work done, the words in *this* language tell us what is real. As our vocabulary changes, so does the world.

Perhaps we can now appreciate what Bohr meant when he said, "But if anybody says he can think about quantum problems without getting giddy, that only shows he has not understood the first thing about them." Physics shows us that while the world shapes us, the language that we use shapes the world. We might even say the language that we *are* shapes the world, for language undoubtedly defines us more profoundly than we can begin to imagine.

EPILOGUE

Setsuji-ichimotsu soku fuchū
"Begin to preach, and the point is lost."

ZENRIN KUSHŪ

The development of physics led Bohr, Einstein, and Heisenberg to the conclusion that, despite how it may seem, physics is not an undistorted picture of an already-made world but a way of talking about the world. The development of the science since the time when these founders of twentieth-century physics did their pioneering work has done nothing to undermine this insight. No slavish reconstruction of a ready-made world, physics is an imaginative vision of how the world *might* be put together.

Although this point is by now obvious, I want to be explicit: Just as there seems to be no "right" way to talk about the world, there is no "right" way to talk about physics, including what I say in this book—that physics is a way of talking about the world. The benefit of using language in this way comes from the advantages the perspective provides. If we find this approach valuable, we will continue to say that physics and the wider world of science and other human creations are ways of talking. If not, we can always choose to ignore Einstein's admonition and continue to see these enterprises, "not as creations of thought, but as given realities."

Is There a Fact in the Matter?

In 1935, Einstein published a paper with Boris Podolsky and Nathan Rosen (EPR) in which they argued that quantum mechanics is incomplete. Their argument is based on the fact that quantum mechanics tells us that there are certain properties of a system, such that when the value of one property is measured with great precision, the value of the other property cannot be determined. As a result, it is impossible to say that the system has particular values of both properties at the same time; an electron cannot be said to have both a definite position and definite momentum at the same time.

Imagine two systems, A and B, that interact and then separate until they are far from each other. Suppose that each system is like a spinning top and that we can measure the direction of the axis about which it is spinning in one of two directions—the axis is either up or down and either right or left. To follow quantum mechanical usage we will say that a measurement of system A will find it either in the up state or in the down state, and another kind of measurement will find A either in the right state or in the left state. Quantum mechanics tells us that A and B can be related in such a way that the following conditions are met: if A is found to be in the up state, then B will be found to be in the down state, and vice versa; also, if *A* is found to be in the left state, then *B* will be found to be in the right state, and vice versa. Furthermore, according to quantum mechanics, if we measure the up-down property it will be impossible to determine if the left-right property was left or right before we measured the up-down property. The same thing is true if we measure the left-right property—we can tell nothing about the up-down property.

Now imagine an experimenter making measurements on one of the systems—say *A*. The experimenter can choose to measure up-down or left-right. Say she chooses up-down and that she finds *A* to be in the down state. Although she has not measured the value for *B*, which is now quite far away, she knows from quantum mechanics that she would find *B* to be in the up state. But she might have measured left-right instead of up-down. Say that if she did, she might have found *A* to be in the left state and would then have known that a measurement of *B* would find it in the right state.

Einstein, Podolsky, and Rosen argued that the experimenter could find a perfectly definite value for *either* *B*'s up-down state or right-left state by measuring *A*'s properties. Since measuring *A*'s properties should not affect *B*'s properties, they reasoned that *B* must have *both* a definite up-down state and a definite right-left state—there must be a fact as to which state *B* is in prior to the measurement of *A*. *B* must be in a definite up-down and right-left state; both must be "elements of reality." However, quantum mechanics tells us we can determine only one or the other. Einstein concluded that quantum mechanics is incomplete—that there are things we know are real, such as the fact about the simultaneously well-defined up-down and right-left states of *B*, that quantum mechanics can tell us nothing about.

Bohr was quick to respond to Einstein's challenge. In fact, Bohr used Einstein's own position to argue against him. Einstein had frequently stated that theories are free creations of the human mind. Bohr argued that the notion of an "element of reality," using Einstein's description, was just such a "free creation" and needed to be checked in the same way that any other theory needed to be checked. After all, everyone had been convinced that it made perfect sense to talk about simultaneous events separated by large distances until Einstein showed that different observers would see different events as simultaneous. In the same way, Bohr argued, we have to pay attention to what we can actually observe rather than what we think physical reality *must* be like.

We can measure the up-down state of A, or we can measure the right-left state of A. If we measure the up-down state of A we can accurately predict the result of measuring the up-down state of B; if we measure the right-left state of A we can accurately predict the result of measuring the right-left state of B. But two different and exclusive measurement of A are called for, and performing either affects the results we get when we then perform the other. Speculation about what we would have learned if we had made measurements that we did not actually make seems to lead nowhere. Measuring A is an indirect measurement of B. We can choose to determine which property of B we will indirectly measure, but we cannot measure both simultaneously and precisely. That is what quantum mechanics tells us, and that is all quantum mechanics tells us.

Let us return to the experiment. Immediately after the interaction of A and B we can imagine two situations: (1) A and B are in definite and opposite up-down and right-left states; (2) A and B are not in definite and opposite up-down and right-left states. Case 1 seems easy enough to understand and may be called the classical situation; it fits our intuitions that there are facts independent of our measurements. But what can Case 2 mean? Case 2 is the quantum mechanical situation. In Case 2, A and B are neither in the right state nor in the left state before we measure them but in some kind of combination we can call left-right state. The up-down state represents a combination of the two possible states up and down before a measurement is made. The situation is exactly analogous to the two-slit experiment. Did the electron go through either slit A or slit B? When we force the electron to go through one slit by closing the other, we get no diffraction pattern. If there is a diffraction pattern, then it seems as if there is no fact about the electron passing through one slit or the other. Rather it is as if the electron passed through *both* slits, even though we never find part of an electron passing through one slit.

Since a measurement always finds A to be in either the right state or the left state, even though we cannot say that A was in either state prior to the measurement, it is sometimes said that a measurement *projects* A from the right-left state into either a right state or a left state. Quantum mechanics predicts that if a measurement of A finds it in the up state, a measurement of B must find it in the down state. This statement seems to imply that when the measurement finds A in the up state, in addition to projecting A into the up state, it also *projects* B into the down state, even though B is nowhere around, since B was also in the up-down state before the measurement of A. When A is measured and found to be in the up state, it seems as if A signals B, "I am in the up state; you go into the down state." This behavior seems to be a strange form of action at a distance. Furthermore, the transmission of physical influence at speeds far faster than light seems to be called for, since A and B can be arbitrarily far apart. The quantum mechanical description violates our deep intuitions that influence must in some sense be local and seems to violate special relativity because it demands that a signal be transmitted faster than light. Exactly the same situation was discussed in terms of the instantaneous collapse of the probability distribution associated with an electron (see page 97).

Case 1 and Case 2 make different predictions about the outcome of the EPR experiment. Once again we can return to the two-slit experiment. If we arrange the experiment so that the electrons must pass through either slit A or slit B (either by closing one slit or placing separate detectors immediately behind each of the two slits), there is no interference pattern. We can state this find in the following way: If we can tell that the electron went through either slit A or slit B, there is no diffraction pattern. But if we cannot tell which slit the electron passed through—if we have to take both slits into account when describing how the electron *might* have gotten to the screen—we *do* get a diffraction pattern. Case 1 is the situation where only one or the other slit is open, and we seemingly can say that there is a fact about which slit the electron passed through. Case 2 is the situation where both slits are open and we seemingly cannot say that there is a fact about which slit the electron passed through. The EPR experiment is like Case 2 and quantum mechanics predicts the correct results—just as it does with the two-slit experiment.

The idea that there is a fact about the paths of electrons needs to be questioned. The action-at-a-distance problem arises because we persist in thinking of a quantum world of independent electrons with classical properties like position and motion and spin-axis orientation. But as Bohr tells us, there is no quantum mechanical world—there is only a quantum mechanical *description* of the world. The "knowing at a distance" arises only if we attempt to explain *how* a subatomic system works. So long as we stick to what we observe, no problems arise—if A is found to be in the up state, B will be found to be in the down state. That is not paradoxical or even difficult to understand. Quantum mechanics requires us to give up the description of separate events and objects on the atomic scale. The detailed behavior of atomic particles is something made up after the fact to "explain" observations.

The Delayed-Choice Two-Slit Experiment

The American physicist John Wheeler drew attention to an example of the two-slit experiment even more confounding than the standard version discussed above. In this arrangement, we place a lens in front of the screen with the two slits and replace the screen that detects the electrons with a shutter. We now direct a stream of photons from a light source toward the apparatus. Behind the shutter we place two photon counters (we could also do the experiment with electrons, as above), so that the image of each slit falls on one counter. When the shutter is closed, the configuration is a simple two-slit experiment, and the photons fall on the shutter in a characteristic light-dark pattern. When the shutter is open, however, the detectors can be used to determine which slit each photon passed through. Thus in the first configuration we observe a diffraction pattern and therefore cannot say that the photon has a unique path that takes it through one slit or the other. In the second configuration we destroy the diffraction pattern, but we can say which slit the photon passed through. The interesting quality about this apparatus is that it is possible to open or close the shutter *after* the photon has passed the slits and before it arrives at the shutter. So after the photon has passed the slits we can apparently reach back in time and force the photon to pass through a single slit or through both!

The delayed two-slit experiment would not have bothered Heisenberg. For him the history we create to explain the arrival of the photon at the counter is a just-so story. Altering the experimental arrangement alters what we can *say* about the paths of the photons. In the delayed-choice experiment we have two different experimental arrangements that we describe in two quite different ways—but the only facts there are about the past histories of photons are stories we invent. What is or is not a fact is determined, not by the world, but by the way we talk about the world. The best model of the subatomic world physicists have will let them say nothing about the simultaneous precisely defined position and motion of a photon.

Schroedinger's Cat, the Problem of Measurement, and Language Domains

Something quite new happened with the advent of quantum mechanics. Relativity could be said to be a replacement for Newtonian mechanics in the sense that we can apply the equations of relativistic mechanics in any case where we can apply the equations of Newtonian mechanics. For example, if the velocities involved are much slower than the velocity of light, as they often are, the relativistic equations become equivalent to the Newtonian equations. In this sense relativistic mechanics includes Newtonian mechanics as a special case. Unlike relativity, quantum mechanics does *not* eliminate the need for nineteenth-century physics. If we attempt to apply the methods of quantum mechanics to the classical domain, we seem to be conjuring up a chimera that we never have to face in reality—the *superposition,* or *overlapping,* of everyday objects.

Schroedinger's Cat

The overlapping of everyday objects was first raised by Erwin Schroedinger in an example that has become quite famous, or infamous, as *Schroedinger's cat*. Schroedinger envisioned a device in which a cat is confined in a box with a flask of poisonous gas. The flask containing the gas will be broken if a radioactive source with a 50-50 chance of emitting an alpha particle during the course of the experiment triggers a detector.

Schroedinger pointed out that before the machine is opened in order to determine whether or not the cat is dead, we can predict the outcome of the experiment using quantum mechanics. The prediction is a combination of the two possible outcomes of the experiment, a blend of a live cat and a dead cat. This blending is completely analogous to the two-slit experiment where we can say that prior to being detected at the second screen, the electron passed through both slits—that it is somehow a combination of passing through the first slit and passing through the second slit. In the case of the electron, this combination is responsible for the interference pattern physicists observe. It is difficult enough for most people to accept an electron in a combination or superposition of states, but a blending of live and dead cats is too much—which was exactly Schroedinger's point. He was pointing to what he saw as a flaw in quantum mechanics.

The Problem of Measurement

Schroedinger's cat is intimately associated with what is sometimes referred to as *the problem of measurement*. The problem comes about because many quantum mechanical calculations lead to probabilities—for example, there may be a number of possible meter readings with a probability assigned to each reading. But we never observe a superposition in the laboratory. We always observe a pointer at one well-defined place. How come? Classical objects such as meters or cats are never found in a superposition of states, but atomic particles are often predicted by quantum mechanics to be in just such superpositions. Yet a classical device is presumably composed of nothing but atomic systems. So why doesn't the everyday world seem to behave the same way atomic systems do?

In the 1930s the Hungarian-born American mathematician John von Neumann analyzed what happens in measuring a quantum mechanical system. He showed that it is possible to place the line separating the quantum mechanical system from the classical laboratory apparatus at a variety of places—that where the line is drawn is to a large degree arbitrary. The location of this line, from our present perspective, simply reflects the point at which physicists stop using one vocabulary to describe the experiment and begin using another. As long as they do not talk about subatomic systems in classical language, the point at which they change vocabularies is one that they choose, not something that nature imposes on them.

The problem of measurement reveals a linguistic gulf between the world of quantum phenomena and the world of everyday experience, including scientific laboratories. If physicists use classical language, such as position and momentum, to describe atomic experiments, they get the wrong answer; it seems as though the electron has to pass through one slit or the other, and we predict no interference. If physicists use quantum mechanical language to talk about classical phenomena, they predict such bizarre things as the blending of live and dead cats. Clearly the quantum mechanical domain and the classical domain require two distinct ways of speaking.

Eugene Wigner argued that since we never find a distribution of meter readings, but always a single reading, and since quantum mechanics seems to predict only probability distributions, human consciousness must somehow induce the probability distribution to collapse. For example, when a human being looks inside Schroedinger's box, he or she causes the probability distribution consisting of a mixture of live and dead cats to collapse and produce either a live cat or a dead cat. We can agree with Wigner that there is something about human beings that leads to this collapse, but it is not necessary to appeal to some nonphysical property of consciousness; instead we can look to our use of language. The collapse of the probability distribution occurs at the point where we feel it is no longer sensible to *talk* about a system in the language of quantum mechanics and we shift to the language of classical physics—when we are unwilling to continue to talk about the superposition of systems. In other words, the collapse of the probability distribution takes place in language.

We encounter domains like those of classical and quantum language in the everyday world. If I want to talk about your behavior, for example, it makes perfect sense to attribute it to a variety of circumstances. I might say, "You did that because you always. . . ." Yet when I am talking about my own behavior, I more often than not attribute it to choices and decisions I make, rather than to external circumstances or the state of my body. These two incompatible ways of talking seem to lie at the heart of the perennial philosophical question of free will and determinism. From the perspective we have been pursuing, such debates look remarkably like arguments about whether objects "really" can be in a superposition of states. As far as physics is concerned, they cannot if they are classical objects like billiard balls and they can if they are subatomic "objects" like electrons. Free will and determinism are nonoverlapping domains like the classical and the quantum mechanical. The best way to become confused is to try to talk about one in the vocabulary of the other. Whenever we fail to distinguish the appropriate domain of a language, we wind up talking about things not very different from the blend of live and dead cats.

NOTES

1. Max Born, *Physics in My Generation* (New York: Springer Verlag, 1969), 49.

2. Albert Einstein, "Johannes Kepler," in *Ideas and Opinions* (New York: Crown, 1954), 266.

3. Nicholas Copernicus, *De Revolutionibus,* quoted in Thomas Kuhn, *The Copernican Revolution: Planetary Astronomy in the Development of Western Thought* (Cambridge: Harvard University Press, 1957), 179–80.

4. Ilse Rosenthal-Schneider, "Reminiscences of Einstein," in *Some Strangeness in the Proportion: A Centennial Symposium to Celebrate the Achievements of Albert Einstein,* ed. Harry Woolf (Reading, Mass.: Addison-Wesley, 1980), 521.

5. Richard Feynman, Robert Leighton, and Matthew Sands, *The Feynman Lectures on Physics,* vol. 1 (Reading, Mass.: Addison Wesley, 1963), 7–2.

6. Galileo Galilei, "The Assayer," in *Discoveries and Opinions of Galileo,* trans. Stillman Drake (New York: Doubleday, 1957), 237–38.

7. James Jeans, *The Mysterious Universe* (Cambridge: Cambridge University Press, 1930), 134.

8. I. Bernard Cohen, *The Birth of a New Physics* (New York: Norton, 1985), 3.

9. When he had finally worked out his model of motion, Newton replaced deductions based on the calculus he had developed with geometric arguments, which were far less elegant but much more familiar to his readers.

10. William Wordsworth, "The Tables Turned—An Evening upon the Same Subject."

11. William Blake, Letter to Thomas Butt, 22 November 1802.

12. Albert Einstein and Leopold Infeld, *The Evolution of Physics* (New York: Simon and Schuster, 1938), 295–96.

13. Feynman et al., vol. 2, 20–29.

14. Richard Feynman, "The Development of the Space-Time View of Quantum Mechanics," *Science* 153 (August 1966): 707–08.

15. Feynman, et al., *The Feynman Lectures on Physics,* vol. 1, 1-2.

16. One formulation of the first law of thermodynamics is "You can't win"; of the second law, "You can't even break even."

17. Albert Einstein, "Time, Space, and Gravitation," in *Out of My Later Years* (New York: Philosophical Library, 1950), 58.

18. Strictly speaking, at this time Einstein talked about the relativity of motion only in systems where gravity was not involved. It was not until he developed the general theory in 1916 that he extended the principles of relativity to include the phenomena of gravity.

19. From the particle's point of view, however, nothing unusual is going on, no matter how fast it is moving with respect to us. If the particle noticed that time was slowing down for it, the particle would know that it was in motion and this would violate Einstein's first principle.

20. Werner Heisenberg, *Physics and Philosophy: The Revolution in Modern Science* (New York: Harper and Row, 1962), 63.

21. Because the strength of gravity changes from point to point, the equivalence of falling freely in a gravity field and floating in space holds strictly only over limited regions of space and time.

22. Heisenberg, *Physics and Philosophy,* 97.

23. John Wheeler, "Hermann Weyl and the Unity of Knowledge," *American Scientist,* 74 (July-August 1986): 371.

24. Max Planck, quoted in Morris Kline, *Mathematics and the Search for Knowledge* (Oxford: Oxford University Press, 1985), 160.

25. Max Planck, "The Genesis and Present State of Development of the Quantum Theory," reprinted in Jefferson Weaver, *The World of Physics: A Small Library of the Literature of Physics from Antiquity to the Present,* vol. 2 (New York: Simon and Schuster, 1987), 284.

26. Clinton Davisson, quoted in Anthony French and Edwin Taylor, *An Introduction to Quantum Physics* (New York: W. W. Norton, 1978), 54.

27. In certain circumstances, this identity breaks down; cf. P. A. M. Dirac, *Lectures on Quantum Field Theory* (New York: Yeshiva University, 1966), 4ff.

28. Max Born, quoted in Abraham Pais, *Inward Bound: Of Matter and Forces in the Physical World* (New York: Oxford University Press, 1986), 256.

29. The probability of finding an electron at some point in space is given by the square of the absolute value of the amplitude of the wave function at that point.

30. In practice, the experimental arrangement would be somewhat different, but the differences are unimportant—the principle remains the same.

31. Victor Weisskopf, *Physics in the Twentieth Century* (Cambridge: MIT Press, 1972), 76.

32. Planck told us that there is energy associated with the frequency of radiation. We will see in the next chapter that Heisenberg uncovered a relationship between energy and time similar to the relationship he described between position and momentum. In order to measure the energy associated with a given frequency, we must take some time to make the measurement—the more accurately we want to know the energy, and hence frequency, the longer it will take to make the measurement. Since the electromagnetic radiation is passing us at the speed of light, the longer it takes to measure its frequency, the longer the wave train must be.

33. P. A. M. Dirac, *The Principles of Quantum Mechanics,* 2d ed. (Oxford: Clarendon Press, 1935), 10.

34. Niels Bohr, quoted in Aage Peterson, "The Philosophy of Niels Bohr," in *Niels Bohr: A Centenary Volume,* eds. A. French and P. Kennedy (Cambridge: Harvard University Press, 1985), 305.

35. Richard Feynman, *The Character of Physical Law* (Cambridge: MIT Press, 1967), 129.

36. Freeman Dyson, *Some Strangeness in the Proportion,* ed. H. Wood (Reading, Mass.: Addison Wesley, 1980), 376.

37. Wolfgang Pauli, quoted in Abraham Pais, *Inward Bound: Of Matter and Forces in the Physical World* (New York: Oxford University Press, 1986), 361.

38. See Feynman's description of this exchange in Crease and Mann, 137–38.

39. Feynman, "The Development of the Space-Time View of Quantum Electrodynamics," 707.

40. Feynman, *QED,* 128.

41. Steven Weinberg, quoted in Crease and Mann, 184.

42. Eugene Wigner, "Invariance in Physical Theory," in *Symmetries and Reflections* (Woodbridge, Conn.: Ox Bow, 1979), 3.

43. Victor Weisskopf, "Quality and Quantity in Quantum Physics," in *Physics in the Twentieth Century: Selected Essays* (Cambridge: MIT Press, 1972), 28.

44. Eugene Wigner, "The Role of Invariance Principles in Natural Philosophy," in *Symmetries and Reflections* (Bloomington: Indiana University Press, 1967), 29.

45. Measurements are still not sufficiently precise to rule out a very small mass for the neutrino.

46. Arthur Eddington, *The Philosophy of Physical Science* (Ann Arbor: University of Michigan Press, 1958), 112.

47. Actually a set of three numbers—one for the component of motion in each of three directions.

48. When equations are written in such a way as to take into account the

requirements of relativity, they are said to be *Lorentz invariant,* after the Dutch physicist Hendrik Lorentz, who first wrote down the equations that Einstein explained in his relativity paper of 1905. Equations that possess Lorentz invariance describe situations in which energy and momentum are conserved. For this reason, physicists say that any set of equations attempting to describe nature must be Lorentz invariant.

49. Since quantum mechanics deals only with probabilities, anything that is not impossible has some probability of occurring. Physicists describe the subatomic world as a freewheeling one where everything that is not prohibited happens at some time or other. We have seen one example of this freewheeling behavior in the virtual processes that physicists say are continually going on in the vacuum.

50. Whenever a particle encounters its antiparticle (an electron encounters a positron, for example), the two particles vanish in a burst of energy. Physicists do not normally refer to this as decay, however. In any case, electric charge and all other quantum numbers are conserved in these processes.

51. The rotation of the nucleon in this abstract space is continuous, so that in addition to heads and tails, all intermediate positions are also possible.

52. Anthony Zee, *Fearful Symmetry: The Search for Beauty in Modern Physics* (New York: Macmillan, 1986), 160.

53. In particular, they wanted to be sure that isospin is conserved.

54. In the approach taken by Yang and Mills, the nucleus of a helium atom contains four nucleons of which two are protons and two are neutrons, but it does not matter which nucleons physicists label neutrons and which they label protons, nor is it necessary that they always label the *same* nucleons protons or neutrons.

55. Freeman Dyson, "Old and New Fashions in Field Theory," *Physics Today* (June 1965): 23.

56. Steven Weinberg, "Conceptual Foundations of the Unified Theory of Weak and Electromagnetic Interactions," *Science* 210 (12 December 1980): 1218.

57. Electrons and neutrinos are placed in a family called *leptons,* the way protons are placed in a family called *baryons.* The weak force affects all the members of this family in a similar fashion.

58. This space is named the *weak isospin* space, following Heisenberg's invention of isospin space.

59. The drama, and melodrama, behind this simple sentence is engagingly told in Gary Taubes, *Nobel Dreams* (New York: Random House, 1986).

60. Albert Einstein, "The Fundaments of Theoretical Physics," *Out of My Later Years* (New York: Philosophical Library, 1950), 98.

61. Computers are taking over particle physics as they are almost everything else. The most modern detectors are computer controlled and computers

determine which data to record for further analysis. The "pictures" these new detectors take are electronic. The principles, however, remain the same.

62. George Zweig, "Origins of the Quark Model," *Baryon '80: Proceedings of the 4th International Conference on Baryon Resonances,* ed. N. Isgur (Toronto: University of Toronto Press, 1981), 439.

63. Like most absolute statements, this one is not completely accurate. Experiments carried out at Stanford University seemed at one time to provide evidence for fractional electric charge, but these results had never been successfully duplicated at other institutions. Physicists therefore discount the original Stanford results.

64. The description of how to transform quarks is different from the description of how to transform leptons because a different symmetry group is involved. Mathematician call the group associated with the electroweak force $SU(2) \times U(1)$ symmetry and the group associated with quarks $SU(3)$.

65. Six gluons transform quarks of one color into another color. Three gluons allow quarks to interact without changing color. When the details of quark-quark interactions are worked out, one of these nine quarks turns out to be redundant, and so eight remain.

66. David Gross, quoted in Crease and Mann, p. 333.

67. Sheldon Glashow, "Towards a Unified Theory; Threads in a Tapestry," *Science* 210 (19 December 1980): 1319.

68. Feynman, *QED,* 15.

69. Steven Weinberg, "The Search for Unity: Notes for a History of Quantum Field Theory," *Daedalus* 106, 2 (1977), 23-33.

70. P. A. M. Dirac, *Lectures on Quantum Field Theory* (New York: Yeshiva University, 1966), 1.

71. Steven Weinberg, quoted in Crease and Mann, 187.

72. Gerard 't Hooft, "Gauge Theories of the Forces between Elementary Particles," *Scientific American* 242, 6 (June 1980): 136.

73. Feynman, *QED,* 139.

74. Richard Feynman and Steven Weinberg, *Elementary Particles and the Laws of Physics: The 1986 Dirac Memorial Lectures* (Cambridge: Cambridge University Press, 1987), 95.

75. L. More, quoted in Bernard Cohen, *Revolution in Science* (Cambridge: Harvard University Press, 1985), 414.

76. Steven Weinberg, quoted in Crease and Mann, 419.

77. Zee, 253-54.

78. In his book Zee's enthusiasm reminds me of Kepler's conviction that he had discerned the plan used by God in fabricating the universe. Time will tell if future generations treat Zee's "new physics" with more reverence than the current generation treats Kepler's "harmony of the spheres."

79. Eugene Wigner, "The Unreasonable Effectiveness of Mathematics in the Natural Sciences," in *Symmetries and Reflections,* 223.

80. Feynman, *QED,* 149.

81. Howard Georgi, quoted in Crease and Mann, 417.

82. It is interesting that the largest facilities built by scientists are needed to test what physicists say about nature on the smallest scales.

83. There does appear to be a net baryon number, because there does not seem to be much antimatter in the universe to balance all the normal matter. But this apparent asymmetry arises in the GUTs out of a completely symmetrical beginning.

84. Alan Guth quoted in M. Waldrop, "The New Inflationary Universe," *Science* 219 (28 January 1983): 375.

85. Eugene Wigner, "The Unreasonable Effectiveness of Mathematics in the Natural Sciences," in *Symmetries and Reflections,* 237.

86. Albert Einstein, "Geometry and Experience," in *Ideas and Opinions,* 233.

87. Morris Kline, *Mathematics: the Loss of Certainty* (Oxford: Oxford University Press, 1980), 322.

88. G. H. Hardy, *A Mathematician's Apology* (Cambridge: Cambridge University Press, 1941), 123–24.

89. Kline, 323.

90. Ibid.

91. Ibid.

92. Ibid.

93. For an illuminating discussion of this question see A. Pickering, *Constructing Quarks: A Sociological History of Particle Physics* (Chicago: University of Chicago Press, 1984).

94. Following the dedication to *Word and Object* (Cambridge: MIT Press, 1960), the American philosopher W. V. O. Quine attributes the aphorism "Ontology recapitulates philology" to James Grier Miller.

95. Albert Einstein, "The Fundaments of Theoretical Physics," 98.

96. Humberto Maturana and Francisco Varela, *The Tree of Knowledge: The Biological Roots of Human Understanding* (Boston: New Science Library, 1987), 22–23.

97. Albert Einstein and Leopold Infeld, *The Evolution of Physics: From Early Concepts to Relativity and Quanta* (New York: Simon and Schuster, 1938), 296.

98. Heisenberg, *Physics and Philosophy,* 58.

99. cf. Richard Rorty, *Philosophy and the Mirror of Nature* (Princeton: Princeton University Press, 1979).

100. Albert Einstein, *The Meaning of Relativity,* 4th ed. (Princeton: Princeton University Press, 1953), 2.

101. Samuel Ting, quoted in Crease and Mann, 366.

102. Explanations such as "the will of God" are compatible with anything that might happen—no evidence could ever conflict with this view, which is why it is not a physical theory.

103. Some equipment, particularly the highest energy accelerator available at any time, may be unique, and so this principle must be relaxed. But it is relaxed only because physicists are convinced that if equally powerful accelerators were built, they would produce similar data.

104. Richard Feynman and Steven Weinberg, *Elementary Particles and the Laws of Physics: The 1986 Dirac Memorial Lectures* (Cambridge: Cambridge University Press, 1987), 95.

105. Einstein, "Physics and Reality," in *Ideas and Opinions,* 292.

106. Ludwig Wittgenstein, *Tractatus Logico Philosophicus,* trans. D. F. Pears and B. F. McGuinness (Atlantic Highlands: Humanities Press, 1974), 74.

107. Einstein and Infeld, 31.

108. Pierre-Simon Laplace, quoted in Kline, 67.

109. J.C. Polkinghorne, *The Quantum World* (Princeton: Princeton University Press, 1985), 38.

110. Ludwig Wittgenstein, *Philosophical Investigations,* trans. G. Anscombe (New York: Macmillan, 1958), 48.

111. Albert Einstein, quoted in Werner Heisenberg, *Physics and Beyond: Encounters and Conversations* (New York: Harper and Row, 1971), 77.

112. Arthur Eddington, *New Pathways in Science* (Ann Arbor: University of Michigan Press, 1959), 21.

113. Niels Bohr, quoted in Aage Peterson, "The Philosophy of Niels Bohr," in *Niels Bohr, A Centenary Volume,* eds. A. French and P. Kennedy (Cambridge: Harvard University Press, 1985), 302.

114. Gilbert Ryle, *The Concept of Mind* (Chicago: University of Chicago Press, 1984), 76.

115. Martin Heidegger, "The Nature of Language," in *On the Way to Language,* trans. Peter Hertz (San Francisco: Harper and Row, 1971), 63.

BIBLIOGRAPHY

Historical

Cohen, Bernard. *The Birth of a New Physics.* New York: W.W. Norton, 1985.

——. *Revolution in Science.* Cambridge: Harvard University Press, 1985.
The evolution of science from the perspective of a respected historian of science.

Crease, Robert, and Charles Mann. *The Second Creation: Makers of the Revolution in 20th-Century Physics.* New York: Macmillan, 1986.
A fascinating history of twentieth-century physics, rich in interviews of the leading physicists of the past 30 years.

Einstein, Albert, and Leopold Infeld. *The Evolution of Physics: From Early Concepts to Relativity and Quanta.* New York: Simon and Schuster, 1938.
A popular introduction by one of the greatest minds of this or any other century.

Holton, Gerald. *Introduction to Concepts and Theories in Physical Science.* Princeton: Princeton University Press, 1985.
A readable and informative historical introduction to the conceptual development of physical science.

——. *Thematic Origins of Scientific Thought: Kepler to Einstein.* Cambridge: Harvard University Press, 1973.
Illuminating essays on the origins of science.

Kline, Morris. *Mathematics: The Loss of Certainty.* New York: Oxford University Press, 1980.
An engaging history of mathematics for the nonmathematical.

——. *Mathematics and the Search for Knowledge.* New York: Oxford University Press, 1985.
A popular discussion of the mathematical aspects of the development of physics.

Pickering, Andrew. *Constructing Quarks: A Sociological History of Particle Physics.* Chicago: University of Chicago Press, 1984.
A fascinating look at physics from the perspective of a sociologist.

Rhodes, Richard. *The Making of the Atomic Bomb*. New York: Simon and Schuster, 1986.
Another very accessible and well-written history of physics during the first half of the century.

Riordan, Michael. *The Hunting of the Quark: A True Story of Modern Physics*. New York: Simon and Schuster, 1987.
The invention of quarks from the viewpoint of an experimental physicist.

Spielberg, Nathan, and Bryon Anderson. *Seven Ideas That Shook the Universe*. New York: John Wiley & Sons, 1987.
Covers much of the same ground as the beginning of this book, albeit from a more traditional perspective and in somewhat more detail, but without becoming overly technical.

Weaver, Jefferson. *The World of Physics: A Small Library of the Literature of Physics from Antiquity to the Present*. 3 vols. New York: Simon and Schuster, 1987.
A collection of many articles, including Nobel acceptance speeches. A goldmine I wish I had had access to when I started writing this book.

Popular Expositions

Davies, Paul. *Superforce*. New York: Simon and Schuster, 1984.

——. *The Forces of Nature*. Cambridge: Cambridge University Press, 1986.
Davies has few peers when it comes to presenting physics in an understandable way.

Feynman, Richard. *The Character of Physical Law*. Cambridge: MIT Press, 1965.
The physicist's physicist at his best—popular lectures that are lucid and illuminating.

——. *QED: The Strange Story of Light and Matter*. Princeton: Princeton University Press, 1985.
The master explains quantum electrodynamics to a nontechnical audience.

Ne'eman, Yuval, and Yoram Kirsh. *The Particle Hunters*. Cambridge: Cambridge University Press, 1986.
A popular treatment by one of the founders of the eightfold way.

Pagels, Heinz. *The Cosmic Code: Quantum Physics as the Language of Nature*. New York. Simon and Schuster, 1982.

——. *Perfect Symmetry: The Search for the Beginning of Time*. New York. Simon and Schuster, 1985.
Pagels writes clearly and has a flair for making technical points understandable to the layperson. The first volume includes an accessible discussion of esoteric interpretations of quantum mechanics.

Trefil, James. *The Moment of Creation: Big Bang Physics from Before the First Millisecond to the Present Universe*. New York: Macmillan, 1983.
Clear and well written.

Weinberg, Steven. *The First Three Minutes: A Modern View of the Origin of the Universe.* New York: Basic Books, 1977.
A masterful, easy-to-read exposition.

Will, Clifford. *Was Einstein Right? Putting General Relativity to the Test.* New York: Basic Books, 1986.
An understandable presentation of relativity and the ways it has been tested.

Zee, Anthony. *Fearful Symmetry; The Search for Beauty in Modern Physics.* New York: Macmillan, 1986.
A clear and accessible approach to modern physics based on symmetry principles.

Philosophical

Davies, Paul. *God and the New Physics.* New York: Simon and Schuster, 1983.
A fine writer of popular treatments of physics extends his range to explore philosophical implications.

Davies, Paul, and J. Brown. *The Ghost in the Atom.* Cambridge: Cambridge University Press, 1986.
Interviews with physicists on the philosophical implications of quantum mechanics.

Eddington, Arthur. *The Philosophy of Physical Science.* Ann Arbor: University of Michigan Press, 1958.
A marvelously well-written exposition of a unique approach to understanding science.

Fine, Arthur. *The Shaky Game: Einstein Realism and the Quantum Theory.* Chicago: University of Chicago Press, 1986.
An illuminating discussion of Einstein's philosophical approach to physics.

French, A., and P. Kennedy. *Niels Bohr: A Centenary Volume.* Cambridge: Harvard University Press, 1985.
A collection of useful articles, including Bohr's illuminating discussion of his exchanges with Einstein.

Heisenberg, Werner. *Physics and Philosophy: The Revolution in Modern Science.* New York: Harper and Row, 1958.

——: *Physics and Beyond: Encounters and Conversations.* New York: Harper and Row, 1971.
One of the founders of quantum mechanics recounts its origins and reflects on its meaning.

Herbert, Nick. *Quantum Reality: Beyond the New Physics.* New York: Doubleday, 1985.
Herbert demonstrates what happens when you ignore Feynman's warning and try to figure out what is going on at the subatomic level.

Morris, Richard. *The Nature of Reality*. New York: McGraw-Hill, 1987.
An accessible discussion of the nature of reality from the perspective of modern physics. The reader may have difficulty keeping the forest in view for the splendor of the trees.

Rorty, Richard. *Philosophy and the Mirror of Nature*. Princeton: Princeton University Press, 1979.
No philosopher of science, Rorty presents a historical philosophical analysis that results in a view of language and reality very similar to the one presented here.

Worf, Benjamin. *Language. Thought and Reality*. Cambridge: MIT Press, 1956.
Essays by the man who articulated the role of language in determining how the world appears.

Somewhat More Technical

Dodd, James. *The Ideas of Particle Physics: An Introduction for Scientists*. Cambridge: Cambridge University Press, 1984.
A technical, yet largely nonmathematical, treatment for scientists.

Jammer, Max. *The Philosophy of Quantum Mechanics*. New York: John Wiley and Sons, 1974.
A detailed discussion of the interpretations of quantum mechanics from a historical perspective.

Pais, Abraham. *Subtle Is the Lord: The Science and Life of Albert Einstein*. New York: Oxford University Press, 1982.

———. *Inward Bound: Of Matter and Forces in the Physical World*. New York: Oxford University Press, 1986.
Pais is uneven, often very technical, but at times he provides lucid descriptions accessible to the nonspecialist. Both volumes are rich in historical detail.

Taylor, Edwin and John Wheeler. *Spacetime Physics*. San Francisco: Freeman, 1966.
Well characterized by its jacket description: "A brief readable exposition of modern relativity theory illustrated and amplified by a wealth of problems, puzzles, and paradoxes and their detailed solutions."

INDEX